CISM COURSES AND LECTURES

The series presents lecture notes, monographs, edited works and proceedings in the field of Mechanics, Engineering, Computer Science and Applied Mathematics.
Purpose of the series in to make known in the international scientific and technical community results obtained in some of the activities organized by CISM, the International Centre for Mechanical Sciences.

INTERNATIONAL CENTRE FOR MECHANICAL SCIENCES

COURSES AND LECTURES - No. 344

STABILITY AND WAVE PROPAGATION IN FLUIDS AND SOLIDS

EDITED BY

G.P. GALDI
UNIVERSITY OF FERRARA

Springer-Verlag Wien GmbH

Le spese di stampa di questo volume sono in parte coperte da
contributi del Consiglio Nazionale delle Ricerche.

This volume contains 20 illustrations

Originally published by CISM, Udine in 1995.

In order to make this volume available as economically and as rapidly as possible the authors' typescripts have been reproduced in their original forms. This method unfortunately has its typographical limitations but it is hoped that they in no way distract the reader.

ISBN 978-3-211-82687-4 ISBN 978-3-7091-3004-9 (eBook)
DOI 10.1007/978-3-7091-3004-9

PREFACE

There is currently a great deal of interest from both the basic scientific and practical engineering point of view in stability and wave propagation in fluids (classical and non-Newtonian) and in solids. The objective of this volume is to emphasize and to encompass the various aspects of interest which include the necessary mathematical analysis background, constitutive theories for materials of differential type, polarized and schock waves, and second sound in solids at low temperature. Specifically, the paper by Drazin is devoted to the stability and bifurcation of Navier-Stokes flow in channels of variable cross-section. The paper by Boulanger and Hayes deals with homogeneous and inhomogeneous wave propagation in elastic media with particular regard to polarized waves. In the paper by Galdi the well-posedness of the problem related to the equations of fluids of grade two is investigated, including a non-linear stability analysis of steady flows. Finally, in the paper by Ruggeri it is considered the relation between hyperbolic balance laws systems and wave propagation, with particular emphasis to schock wave structures.

These papers are the content of a series of Lectures delivered by the above mentioned authors at the Advanced School "Stability and Wave Propagation in Fluids and Solids" held at the International Center for Mechanical Sciences (CISM) in Udine, during the period of may 24-28, 1993. The Course has been attended by more than fourty participants coming from different Countries. I would like to take this opportunity to thank all of them for the stimulating discussions, and for the lively atmosphere they set up during and after the lectures.

Last but not the least, it is my pleasure as organizer of the School and, therefore, on behalf of all participants, to thank Professor G. Bianchi, Secretary General of the CISM, for his warm hospitality and unselfish help.

G.P. Galdi

CONTENTS

WAVES IN ELASTIC MEDIA

Ph. Boulanger
Free University of Brussels, Brussels, Belgium
and
M. Hayes
University College Dublin, Dublin, Ireland

1. Introduction

We consider the propagation of plane waves in the context of linearized elasticity theory. Both homogeneous and inhomogeneous waves are considered. Propagation in internally constrained media is briefly considered as well as propagation in unconstrained anisotropic and isotropic media.

Modern work on wave propagation in elastic crystals was initiated by Musgrave (1955). He drew attention to the early work by Kelvin (1904) and through a series of papers on theoretical and experimental aspects provided the foundation for later developments. He established that there is very good agreement between theory and experiment.

For homogeneous plane waves a direction of propagation is prescribed and then an algebraic eigenvalue problem is solved for the determination of the corresponding amplitudes and speeds of propagation. We consider the energy flux for such waves.

For inhomogeneous plane waves the planes of constant phase are not in general parallel to the planes of constant amplitude. Before treating these we briefly consider the theory of bivectors or complex vectors. The link between bivectors and inhomogeneous plane waves is that the field, for example, the displacement $\boldsymbol{u}$, corresponding to a train of such waves is described in terms of two bivectors. Thus $\boldsymbol{u} = [\boldsymbol{A} \exp i\omega(\boldsymbol{S} \cdot \boldsymbol{x} - t)]^{+}$. Here ω is real and the amplitude bivector $\boldsymbol{A}$ determines the plane of polarization. The slowness bivector $\boldsymbol{S} = \boldsymbol{S}^{+} + i\boldsymbol{S}^{-}$ determines the planes of constant phase $\boldsymbol{S}^{+} \cdot \boldsymbol{x} =$ constant, the planes of constant amplitude $\boldsymbol{S}^{-} \cdot \boldsymbol{x} =$ constant, the phase slowness $|\boldsymbol{S}^{+}|$ and the attenuation factor $|\boldsymbol{S}^{-}|$. The superscripts + and − are used to denote the real and imaginary parts of a complex quantity.

For inhomogeneous plane waves the directions of $\boldsymbol{S}^+$ and of $\boldsymbol{S}^-$ may not be chosen arbitrarily. For example, in an isotropic elastic material $\boldsymbol{S}^+ \cdot \boldsymbol{S}^- = 0$, whereas for an isotropic viscoelastic material $\boldsymbol{S}^+ \cdot \boldsymbol{S}^- \neq 0$. It is seen (Hayes, 1984) that if $\boldsymbol{S}$ is written $\boldsymbol{S} = N\boldsymbol{C}$, where N is a complex scalar and $\boldsymbol{C} = m\widehat{\boldsymbol{m}} + i\widehat{\boldsymbol{n}}$ is a prescribed bivector with $m, \widehat{\boldsymbol{m}}$ and $\widehat{\boldsymbol{n}}$ prescribed, where m is real and $\widehat{\boldsymbol{m}}, \widehat{\boldsymbol{n}}$ are two orthogonal unit vectors, then N is determined from an algebraic eigenvalue problem. Also, the corresponding amplitude bivectors $\boldsymbol{A}$ are the corresponding eigenbivectors.

We consider especially circularly polarised waves - both homogeneous and inhomogeneous.

We also examine the energy flux for homogeneous and inhomogeneous wave trains from a general standpoint.

2. Basic Equations

Referred to a rectangular cartesian coordinate system $Oxyz$, the particle displacement components are denoted by (u, v, w) or (u_i), and the strains are given by

$$2e_{ij} = \partial u_i/\partial x_j + \partial u_j/\partial x_i \equiv u_{i,j} + u_{j,i} \; . \qquad (2.1)$$

The displacement components $u_i(x_k, t)$ are always assumed to exist and to be real continuously differentiable functions of x_k and t.

The traction $\boldsymbol{t}_{(\boldsymbol{n})}$ across a material element, with unit normal $\boldsymbol{n}$, has components

$$t_{(\boldsymbol{n})_i} = t_{ij} n_j \; , \qquad (2.2)$$

where t_{ij} are the components of the stress tensor which is symmetric. In the absence of body forces the equations of motion are

$$t_{ij,j} = \rho \partial^2 u_i / \partial t^2 \; , \qquad (2.3)$$

where ρ is the material density.

The constitutive equations for a general anisotropic homogeneous elastic material are

$$t_{ij} = c_{ijkl} e_{kl} \; , \qquad (2.4)$$

where c_{ijkl} are constants with the symmetries

$$c_{ijkl} = c_{jikl} = c_{ijlk} = c_{klij} \, . \qquad (2.5)$$

Because of these symmetries there are at most twenty-one independent elastic constants.

Inserting (2.4) into (2.3) and using (2.1) leads to

$$c_{ijkl}u_{k,lj} = \rho\partial^2 u_i/\partial t^2 \, . \tag{2.6}$$

The corresponding equilibrium equations, for static deformations, are

$$c_{ijkl}u_{k,lj} = 0 \, . \tag{2.7}$$

We digress briefly for reference purposes to consider various classifications of the system (2.7). These classifications are considered in the context of the uniqueness of solution for the Dirichlet problem in which (2.7) are to be solved for $\boldsymbol{u}$ in a bounded region B, subject to $\boldsymbol{u}$ being prescribed on ∂B, the boundary of B.

The equilibrium equations are said to be "elliptic" in the sense of Petrovsky if

$$\det(c_{ijkl}a_l a_j) \neq 0 \quad \forall \boldsymbol{a} \neq 0 \, , \tag{2.8}$$

and to be "strongly-elliptic " (S - E) if (Browder 1954, Morrey 1954)

$$c_{ijkl}a_i a_k b_l b_j > 0 \, , \quad \forall \boldsymbol{a}, \boldsymbol{b} \neq 0 \, . \tag{2.9}$$

The equilibrium equations are said to be modified moderately-strongly elliptic (MMSE) if (Hayes 1979) there exists a set of constants ϕ_{ij} with $\det \phi \neq 0$ such that

(i) $\phi_{pi}c_{ijkl}a_p a_k b_l b_j \geq 0$;

(ii) $\phi_{pi}c_{ijkl}a_p a_k b_l b_j = 0$ implies $a_i = 0$ except on a set of zero measure (2.10)

in $\boldsymbol{b}$- space.

The ellipticity condition (2.8) of Petrovsky is a natural generalisation to a system of the ellipticity condition for a single second order partial differential equation, and ensures analyticity of solutions. However it does not guarantee uniqueness of solution of the Dirichlet problem for the system (2.7). The S-E ellipticity condition (2.9) implies the ellipticity condition (2.8) and has close links with wave propagation as will be seen later (§5). However, whilst the S-E condition guarantees uniqueness of solution for the Dirichlet problem for the system (2.7), there are many examples (Hayes, 1963) in which there is uniqueness of solution but the S-E condition is not satisfied. The MMSE condition (2.10) is necessary and sufficient for uniqueness of solutions for the Dirichlet problem for the system (2.7) in the case of cubic crystals (Hayes, 1979).

We assume throughout that the S-E condition (2.9) is satisfied.

3. Special Materials

Here we list the constitutive equations and equations of motion for various material symmetries.

For **isotropic materials** the constitutive equations are

$$t_{ij} = 2\mu e_{ij} + \lambda e_{kk}\delta_{ij} \ , \tag{3.1}$$

where μ (the shear modulus) and λ (Lame constant) are constants. Inserting (3.1) into (2.3) leads to the equations of motion

$$\mu u_{i,jj} + (\lambda + \mu)u_{j,ij} = \rho\partial^2 u_i/\partial t^2 \ . \tag{3.2}$$

For **transversely isotropic materials** the constitutive equations are (Love, 1927)

$$\begin{aligned} t_{11} &= d_{11}u_{1,1} + d_{12}u_{2,2} + d_{13}u_{3,3} \ , \quad t_{12} = \{(d_{11} - d_{12})/2\}(u_{1,2} + u_{2,1}) \ , \\ t_{22} &= d_{12}u_{1,1} + d_{11}u_{2,2} + d_{13}u_{3,3} \ , \quad t_{13} = d_{44}(u_{1,3} + u_{3,1}) \ , \\ t_{33} &= d_{13}(u_{1,1} + u_{2,2}) + d_{33}u_{3,3} \ , \quad t_{23} = d_{44}(u_{2,3} + u_{3,2}) \ , \end{aligned} \tag{3.3}$$

where $d_{11}, d_{12}, d_{13}, d_{33}, d_{44}$ are five material constants. These equations also describe linear elastic crystals with hexagonal symmetry (Fedorov, 1968). The equations of motion now read

$$\begin{aligned} &d_{11}u_{1,11} + \{(d_{11} - d_{12})/2\}u_{1,22} + d_{44}u_{1,33} + \{(d_{11} + d_{12})/2\}u_{2,12} \\ &\qquad + (d_{13} + d_{44})u_{3,13} = \rho\partial^2 u_1/\partial t^2 \ , \\ &\{(d_{11} + d_{12})/2\}u_{1,12} + \{(d_{11} - d_{12})/2\}u_{2,11} + d_{11}u_{2,22} + d_{44}u_{2,33} \\ &\qquad + (d_{13} + d_{44})u_{3,23} = \rho\partial^2 u_2/\partial t^2 \ , \\ &(d_{13} + d_{44})(u_{1,13} + u_{2,23}) + d_{44}(u_{3,11} + u_{3,22}) + d_{33}u_{3,33} = \rho\partial^2 u_3/\partial t^2 \ . \end{aligned} \tag{3.4}$$

For **elastic cubic crystals** the stresses are given by (Love 1927)

$$\begin{aligned} t_{11} &= c_{11}u_{1,1} + c_{12}(u_{2,2} + u_{3,3}) \ , \quad t_{12} = c_{44}(u_{1,2} + u_{2,1}) \ , \\ t_{22} &= c_{12}u_{1,1} + c_{11}u_{2,2} + c_{12}u_{3,3} \ , \quad t_{23} = c_{44}(u_{2,3} + u_{3,2}) \ , \\ t_{33} &= c_{12}(u_{1,1} + u_{2,2}) + c_{11}u_{3,3} \ , \quad t_{31} = c_{44}(u_{1,3} + u_{3,1}) \ , \end{aligned} \tag{3.5}$$

and the equations of motion are

$$\begin{aligned} c_{11}u_{1,11} + c_{44}u_{1,22} + c_{44}u_{1,33} + (c_{12} + c_{44})(u_{2,12} + u_{3,13}) &= \rho\partial^2 u_1/\partial t^2 \ , \\ (c_{12} + c_{44})u_{1,12} + c_{44}u_{2,11} + c_{11}u_{2,22} + c_{44}u_{2,33} + (c_{12} + c_{44})u_{3,23} &= \rho\partial^2 u_2/\partial t^2 \ , \\ (c_{12} + c_{44})(u_{1,13} + u_{2,23}) + c_{44}u_{3,11} + c_{44}u_{3,22} + c_{11}u_{3,33} &= \rho\partial^2 u_3/\partial t^2 \ . \end{aligned} \tag{3.6}$$

For a homogeneous **material inextensible in the direction *l*** the condition of inextensibility is

$$e_{ij}l_i l_j = 0\ . \tag{3.7}$$

The constitutive equations are

$$t_{ij} = Tl_i l_j + d_{ijkl}e_{ke}\ , \quad e_{ij}l_i l_j = 0\ , \tag{3.8}$$

where $T = T(\boldsymbol{x}, t)$ is a tension to be determined from the equations of motion and boundary and initial conditions and $\boldsymbol{d}$ is a fourth order constant tensor with the same symmetries as $\boldsymbol{c}$ has in (2.5). The equations of motion are

$$T_{,j}l_i l_j + d_{ijkl}u_{k,lj} = \rho\partial^2 u_i/\partial t^2\ . \tag{3.9}$$

Finally, the constitutive equations for a homogeneous **anisotropic incompressible elastic material** are

$$t_{ij} = -p\delta_{ij} + d_{ijkl}e_{kl}\ , \quad e_{kk} = 0\ , \tag{3.10}$$

where $p = p(\boldsymbol{x}, t)$ is a hydrostatic pressure to be determined from the equations of motion and boundary conditions, and d_{ijkl} are constants with the same symmetries as c_{ijkl} has in (2.5). The equations of motion are

$$-p_{,i} + d_{ijkl}u_{k,lj} = \rho\partial^2 u_i/\partial t^2\ . \tag{3.11}$$

Materials inextensible in the direction $\boldsymbol{l}$ may be used to model infinitesimal deformations of materials which are reinforced with parallel inextensible fibres in the direction $\boldsymbol{l}$. The constitutive equations (3.10) may be used to model infinitesimal deformations of materials such as rubber in which there are no changes in material volumes during deformation.

4. Waves in Isotropic Media

We consider the propagation of homogeneous plane waves in a homogeneous isotropic elastic material.

Let the displacement components u_i be given by

$$u_i = \{A_i \exp ik(\boldsymbol{n}\cdot\boldsymbol{x} - ct)\}^+\ , \tag{4.1}$$

where $\boldsymbol{A}$ is a vector, possibly complex, k is a real scalar, $\boldsymbol{n}$ is a unit vector and c is a constant. This describes an infinite train of homogeneous plane waves propagating in the direction $\boldsymbol{n}$ with speed c, wavelength $(2\pi/k)$, angular frequency $\omega = kc$, and period $2\pi/(kc)$. Inserting (4.1) into the equations of motion (3.2) leads to the propagation condition

$$q_{ik}(\boldsymbol{n})A_k = \rho c^2 A_i\ , \tag{4.2}$$

where

$$q_{ik} = (\lambda + \mu)n_i n_k + \mu n_j n_j \delta_{ik}$$
$$= (\lambda + 2\mu)n_i n_k + \mu(m_i m_k + p_i p_k) \; . \tag{4.3}$$

Here $\boldsymbol{m}$ and $\boldsymbol{p}$ are unit vectors forming an orthonormal triad with $\boldsymbol{n}$ so that $\delta_{ik} = n_i n_k + m_i m_k + p_i p_k$. Thus the propagation condition may be written

$$[\{(\lambda + 2\mu) - \rho c^2\}n_i n_k + \{\mu - \rho c^2\}(m_i m_k + p_i p_k)]A_k = 0 \; . \tag{4.4}$$

The secular equation is

$$\det(\boldsymbol{q}(\boldsymbol{n}) - \rho c^2 \mathbf{1}) = 0 \; , \tag{4.5}$$

and using the form (4.4) of (4.2), it leads to

$$(\lambda + 2\mu - \rho c^2)(\mu - \rho c^2)^2 = 0 \; . \tag{4.6}$$

Thus, the secular equation has the simple root

$$\rho c_p^2 = \lambda + 2\mu \; , \tag{4.7}$$

and the double root

$$\rho c_s^2 = \mu \; . \tag{4.8}$$

Corresponding to the simple root, it is immediately seen from the propagation condition (4.4) that

$$\boldsymbol{m}(\boldsymbol{A} \cdot \boldsymbol{m}) + \boldsymbol{p}(\boldsymbol{A} \cdot \boldsymbol{p}) = 0 \; ,$$

and hence $\boldsymbol{A} \| \boldsymbol{n}$. Thus, the corresponding expression for the displacement is

$$\boldsymbol{u} = \boldsymbol{n}\{\alpha \exp ik(\boldsymbol{n} \cdot \boldsymbol{x} - c_p t)\}^+ \; , \tag{4.9}$$

where α is an arbitrary (complex) constant, and $\{ \}^+$ means that the real part is being taken of the expression within brackets. The wave is longitudinal - sometimes called a "P - wave" where "P" stands for Push-Pull because it is longitudinal, or Primary because it is the first wave to arrive from a distant earthquake source.

Corresponding to the double root (4.8), it is immediately seen that if we put $\rho c^2 = \mu$ in the propagation condition (4.4), then $\boldsymbol{n}(\boldsymbol{A} \cdot \boldsymbol{n}) = 0$ and hence $\boldsymbol{A} \perp \boldsymbol{n}$. Thus $\boldsymbol{A} = \beta \boldsymbol{m} + \gamma \boldsymbol{p}$, where β, γ are arbitrary (complex) constants. The corresponding displacement is

$$\boldsymbol{u} = \{(\beta \boldsymbol{m} + \gamma \boldsymbol{p}) \exp ik(\boldsymbol{n} \cdot \boldsymbol{x} - c_s t)\}^+ \; . \tag{4.10}$$

This wave is transverse, sometimes called an "S - wave" where "S" refers to Shear-Shake because it is transverse, or Secondary because it is the second to arrive from a distant earthquake source.

Note that for the longitudinal wave (4.9) the displacement is linear, so that the wave is linearly polarized.

However for the shear wave (4.10), the wave is elliptically polarised, in general, for (β/γ) complex; when (β/γ) is real it is linearly polarised and when $\beta/\gamma = \pm i$ it is circularly polarized. Indeed, suppose for simplicity that we have $\beta = 1$ and $\gamma = \gamma^+ + i\gamma^-$. Then

$$\boldsymbol{u} = (\boldsymbol{m} + \gamma^+\boldsymbol{p}) \cos k(\boldsymbol{n}\cdot\boldsymbol{x} - c_s t) - \gamma^-\boldsymbol{p} \sin k(\boldsymbol{n}\cdot\boldsymbol{x} - c_s t) , \tag{4.11}$$

so that the particle initially at $\boldsymbol{x}$ is displaced to $\boldsymbol{x} + \boldsymbol{u}$, and moves on an ellipse, a pair of whose conjugate radii are $\boldsymbol{m} + \gamma^+\boldsymbol{p}$ and $\gamma^-\boldsymbol{p}$. If $\gamma^+ = 0$, $\gamma^- = 1$, then

$$\boldsymbol{u} = \boldsymbol{m} \cos k(\boldsymbol{n}\cdot\boldsymbol{x} - c_s t) - \boldsymbol{p} \sin k(\boldsymbol{n}\cdot\boldsymbol{x} - c_s t) , \tag{4.12}$$

so that the particle moves on a circle. If $\gamma^- = 0$, the wave is clearly linearly polarised along $\boldsymbol{m} + \gamma^+\boldsymbol{p}$.

5. Waves in Crystals

If the expression (4.1) for the displacement is inserted into (2.6), it leads to the propagation condition

$$(q_{ik}(\boldsymbol{n}) - \rho c^2 \delta_{ik}) A_k = 0 , \tag{5.1}$$

and the secular equation

$$\det(\boldsymbol{q}(\boldsymbol{n}) - \rho c^2 \mathbf{1}) = 0 , \tag{5.2}$$

where $\boldsymbol{q}(\boldsymbol{n})$, called the "acoustical tensor" , is given by

$$q_{ik}(\boldsymbol{n}) = c_{ijkl} n_j n_l . \tag{5.3}$$

The secular equation (5.2) is a cubic in ρc^2. Using the symmetry relations (2.5) it follows that $\boldsymbol{q}(\boldsymbol{n})$ is symmetric. From the fact that $\boldsymbol{q}(\boldsymbol{n})$ is real and symmetric, it follows that the three roots, denoted by $c_\alpha^2 (\alpha = 1,2,3)$, of the secular equation are all real, and also that the corresponding eigenbivectors denoted by $\boldsymbol{A}^{(\alpha)} (\alpha = 1,2,3)$ are mutually orthogonal:

$$\boldsymbol{A}^{(\alpha)} \cdot \boldsymbol{A}^{(\beta)} = 0 , \quad \alpha \neq \beta, \quad \alpha, \beta = 1,2,3 . \tag{5.4}$$

If there are no repeated roots, the eigenbivectors $\boldsymbol{A}^{(\alpha)}$ have real directions, that is they are (possibly complex) scalar multiples of real vectors. Thus $\boldsymbol{A}^{(1)} = \lambda_1 \boldsymbol{a}^{(1)}, \boldsymbol{A}^{(2)} = \lambda_2 \boldsymbol{a}^{(2)}, \boldsymbol{A}^{(3)} = \lambda_3 \boldsymbol{a}^{(3)}$ where $\boldsymbol{a}^{(\alpha)}$ are real vectors and λ_α are scalar factors, possibly complex. Indeed, the components of $\boldsymbol{A}^{(\alpha)}$ may be taken to be proportional to the cofactors of the elements of any row of the real matrix $\{q_{ik}(\boldsymbol{n}) - \rho c_\alpha^2 \delta_{ik}\}$.

Hence, corresponding to the direction of propagation $\boldsymbol{n}$, there are, in general, three linearly polarised homogeneous plane wave solutions, the amplitude vectors being mutually orthogonal. For a given wave-length $2\pi/k$, these are

$$\boldsymbol{u}^{(\alpha)} = \{\boldsymbol{a}^{(\alpha)} \exp ik(\boldsymbol{n}\cdot\boldsymbol{x} \pm c_\alpha t)\}^+ , \quad \alpha = 1,2,3, \text{ no sum} . \tag{5.5}$$

Alternatively, for a given angular frequency ω,

$$\boldsymbol{u}^{(\alpha)} = \{\boldsymbol{a}^{(\alpha)} \exp i\omega(\pm c_\alpha^{-1}\boldsymbol{n}\cdot\boldsymbol{x} - t)\}^+ , \quad \alpha = 1,2,3, \text{ no sum} . \tag{5.6}$$

Remark 1. Universal Relations

From (5.2) it follows that the sum of the squared speeds, $c_\alpha^2(\boldsymbol{n})$, of the waves propagating along $\boldsymbol{n}$, is given by

$$\rho \sum_{\alpha=1}^{3} c_\alpha^2(\boldsymbol{n}) = \text{ trace } \boldsymbol{q}(\boldsymbol{n}) = c_{ijil} n_j n_l . \tag{5.7}$$

It follows that if $\boldsymbol{n}, \boldsymbol{m}, \boldsymbol{p}$ form an orthonormal triad, then (Hayes, 1972)

$$\rho\Sigma[c_\alpha^2(\boldsymbol{n}) + c_\alpha^2(\boldsymbol{m}) + c_\alpha^2(\boldsymbol{p})] = c_{ijij} , \tag{5.8}$$

and is thus invariant for the material under consideration.

Remark 2. S-E condition

Using the expression (5.5), for a wave propagating in the direction $\boldsymbol{n}$ with speed c_α and real amplitude vector $\boldsymbol{a}^{(\alpha)}$, it follows that

$$c_{ijkl} a_k^{(\alpha)} n_l n_j = \rho c_\alpha^2 a_i^{(\alpha)} , \tag{5.9}$$

and hence

$$\rho c_\alpha^2 a_i^{(\alpha)} a_i^{(\alpha)} = c_{ijkl} a_i^{(\alpha)} a_k^{(\alpha)} n_l n_j . \tag{5.10}$$

Thus if the S-E condition (2.9) is satisfied, it follows that $c_\alpha^2 > 0$. It has been seen that c_α^2 are real when $\boldsymbol{q}$ is real and symmetric. The S-E condition ensures that c_α are real.

If the S-E condition holds so that c_α are real, then also there is uniqueness of solution, to the Dirichlet problem. However, as remarked in §2, there may be uniqueness of solution to Dirichlet problem when the S-E condition is not satisfied, and the wave speeds are not real. (see Hayes 1963).

Remark 3. Slowness Surface

On a line along $\boldsymbol{n}$, through a given origin 0, the values of the three slownesses $c_1^{-1}, c_2^{-1}, c_3^{-1}$, corresponding to the direction $\boldsymbol{n}$, may be marked off. Then, as $\boldsymbol{n}$ varies over the unit sphere, the corresponding points fill out three closed surfaces, forming a surface of three sheets called the "slowness surface" . Every vector $\boldsymbol{s} = (1/c)\boldsymbol{n}$ having 0 as origin and a point on the surface as extremity, is a possible slowness vector. Because the wave-speeds are obtained from the secular equation (5.2) and because $\boldsymbol{q}(\boldsymbol{n})$ is quadratic in $\boldsymbol{n}$, the slowness surface has for equation

$$\det(\boldsymbol{q}(\boldsymbol{s}) - \rho\mathbf{1}) = 0 \,, \tag{5.11}$$

where $\boldsymbol{s} = (1/c)\boldsymbol{n}$ denotes the slowness vector.

In order to write separate equations for the three sheets of the slowness surface (5.11), the secular equation (5.2) has to be solved for c^2, yielding the three roots $c = V_1(\boldsymbol{n}), c = V_2(\boldsymbol{n}), c = V_3(\boldsymbol{n})$ (say) where c denotes the positive square root of c^2. Because $\boldsymbol{q}(\boldsymbol{n})$ is quadratic in $\boldsymbol{n}$, it follows that V_1, V_2, V_3 are homogeneous functions of degree one in $\boldsymbol{n}$. Thus, because $\boldsymbol{s} = (1/c)\boldsymbol{n}$, the three sheets of the slowness surface have for equations

$$V_\alpha(\boldsymbol{s}) = 1 \,, \quad \alpha = 1,2,3 \,. \tag{5.12}$$

Also, $c = \omega/k$ and thus $\boldsymbol{s} = (1/\omega)\boldsymbol{k}$ and hence the dispersion relations corresponding to the three waves are

$$\omega = V_\alpha(\boldsymbol{k}) \,, \quad \alpha = 1,2,3 \,. \tag{5.13}$$

Remark 4. Double Roots

It will be seen later (§17) that if the secular equation has a double root for some $\boldsymbol{n}$, then a circularly polarised wave of either handedness may propagate along $\boldsymbol{n}$, and thus also elliptically polarised waves may propagate along $\boldsymbol{n}$. Conversely if a circularly polarised wave may propagate along a direction $\boldsymbol{n}$, then the corresponding secular equation must have at least a double root.

If the secular equation has a double root for some $\boldsymbol{n}$, then the sheets of the slowness surface touch or intersect for that $\boldsymbol{n}$. Thus one can tell at a glance at the slowness surface whether or not circularly polarised waves are possible for a given direction.

6. Energy Flux

Because waves carry energy it is useful to have expressions for their energy flux and energy density.

Let $r\Delta a$ be the rate at which energy crosses the element of area Δa of a material surface in the direction of the unit outward normal $\boldsymbol{n}$. This is equal to negative of the rate at which work is being done by forces acting on the surface. Thus

$$r\Delta a = -t_{(\boldsymbol{n})i} v_i \Delta a = -t_{ij} n_j v_i \Delta a = r_j n_j \Delta a \ , \tag{6.1}$$

where

$$r_j = -t_{ij} v_i \ , \tag{6.2}$$

and $v_i = \partial u_i/\partial t$ is the particle velocity.

Taking $n_j = \delta_{j1}$, we see that r_1 is the rate at which mechanical energy crosses at time t, a material element which is normal to the x_1 - axis, measured per unit area of surface at time t. Analogous meanings attach to r_2 and r_3. $\boldsymbol{r}$ is called the "energy flux vector" .

From the equations of motion (3.3) we obtain

$$(t_{ij} v_i)_{,j} - t_{ij} v_{i,j} = \partial K/\partial t \ , \tag{6.3}$$

where

$$K = (1/2)\rho v_i v_i \ , \tag{6.4}$$

is the kinetic energy density. Using the symmetry of the stress tensor, the consitutive equation (2.4), and (2.5), we have

$$\begin{aligned} t_{ij}\partial(\partial u_i/\partial x_j)/\partial t = t_{ij}\partial e_{ij}/\partial t &= c_{ijkl}[\partial e_{ij}/\partial t]\ e_{kl} \\ &= (1/2)\partial(t_{ij}e_{ij})/\partial t \ . \end{aligned} \tag{6.5}$$

Hence we obtain the energy conservation equation

$$\partial e/\partial t + \nabla \cdot \boldsymbol{r} = 0 \ , \tag{6.6}$$

where

$$e = K + \sigma \ , \quad 2\sigma = t_{ij} e_{ij} \ . \tag{6.7}$$

Here e is the total energy density and is the sum of the kinetic energy density K and the elastic stored energy density σ.

For waves with period $\tau \equiv 2\pi/\omega$ we use a tilde to denote means taken over a cycle. Thus the mean energy flux $\tilde{\boldsymbol{r}}$ and the mean energy density $\tilde{e}$ given by

$$\tilde{\boldsymbol{r}} = (1/\tau)\int_0^\tau \boldsymbol{r}\,dt \ , \quad \tilde{e} = (1/\tau)\int_0^\tau e\,dt \ . \tag{6.8}$$

For example, for the wave train (4.1) propagating in an elastic crystal, we find

$$\widetilde{K} = (1/4)\rho\omega^2 \boldsymbol{A}\cdot\overline{\boldsymbol{A}} \ , \quad \tilde{\sigma} = (1/4)k^2 c_{ijkl} A_i \overline{A}_k n_j n_l \ . \tag{6.9}$$

On using the propagation condition (5.1) it follows that $\widetilde{K} = \widetilde{\sigma}$, so that for the total mean energy density $\widetilde{e}$, we have $\widetilde{e} = 2\widetilde{K} = 2\sigma$. Introducing (4.1) into (6.2), and taking the mean over a cycle gives

$$\widetilde{r}_i = (1/4)k^2 c_{ijkl}(A_j\overline{A}_k + \overline{A}_j A_k)n_l \ . \tag{6.10}$$

Using (5.1), we note that

$$\widetilde{\boldsymbol{r}} \cdot \boldsymbol{n} = c\widetilde{e} \ . \tag{6.11}$$

Thus the projection of the mean energy flux vector onto the propagation direction is equal to the product of the wave speed by the mean energy density.

7. Group Velocity for trains of homogeneous plane waves

Here we attempt to determine the general *form* of the energy flux vector for linear conservative systems. Elastic media whether anisotropic or not, whether internally constrained or not, are examples of such linear conservative systems. The purpose in being as general as possible is that it enables us to see the broad picture and to obtain general results without going into the fine details.

We consider the propagation of a single infinite train of elliptically polarised time-harmonic homogeneous plane waves propagating in a homogeneous linear conservative dispersive system.

We make three assumptions.

(1) The energy flux vector $\boldsymbol{r}$, and the energy density e, involve products of pairs of field quantities.

(2) There is neither internal energy supply nor dissipation.

(3) The system is linear in the sense that if one field quantity such as displacement, velocity or stress, for example, is of the form $\boldsymbol{A} \exp i(\boldsymbol{k} \cdot \boldsymbol{x} - \omega t)$, then every other field quantity is of the similar form $\boldsymbol{B} \exp i(\boldsymbol{k} \cdot \boldsymbol{x} - \omega t)$. Here $\boldsymbol{A}$ and $\boldsymbol{B}$ may be (complex) scalars, vectors or tensors.

Suppose now that for the system under consideration the dispersion relation is

$$\omega = f(\boldsymbol{k}), \tag{7.1}$$

where f is a known function. If ω and $\boldsymbol{k}$ satisfy (7.1), then a wave train represented by

$$\boldsymbol{C} \exp i(\boldsymbol{k} \cdot \boldsymbol{x} - \omega t) \ , \tag{7.2}$$

may propagate in the system. Here $\boldsymbol{C}$ is a constant scalar, vector or tensor, in general complex. Due to assumption (1) the energy flux vector associated with (7.2) must have the form

$$\boldsymbol{r} = \boldsymbol{A} \exp 2i(\boldsymbol{k} \cdot \boldsymbol{x} - \omega t) + \overline{\boldsymbol{A}} \exp -2i(\boldsymbol{k} \cdot \boldsymbol{x} - \omega t) + \boldsymbol{w} \ , \tag{7.3}$$

where $\boldsymbol{w}$ is a real constant vector and $\boldsymbol{A}$ is a constant bivector or complex vector. These vectors are functions of ω and $\boldsymbol{k}$, and of the parameters specifying the system. The reason it must have this form is that it is a product of the field quantities both of the form (7.2) and in determining the real vector $\boldsymbol{r}$, the real parts must be taken in each of the expressions in the product. Clearly, using (6.8), $\tilde{\boldsymbol{r}} = \boldsymbol{w}$.

The energy density e associated with the wave train, has a form similar to that of $\boldsymbol{r}$. Thus

$$e = A \exp\{2i(\boldsymbol{k} \cdot \boldsymbol{x} - \omega t)\} + \overline{A} \exp\{-2i(\boldsymbol{k} \cdot \boldsymbol{x} - \omega t)\} + \gamma \ , \tag{7.4}$$

where γ is a real constant and A is a complex constant. The constants are functions of ω and $\boldsymbol{k}$ and of the parameters specifying the system. Clearly $\tilde{e} = \gamma$.

The energy flux velocity may be defined as

$$\boldsymbol{g} = \frac{\text{mean energy flux}}{\text{mean energy density}} = \frac{\tilde{\boldsymbol{r}}}{\tilde{e}} \ . \tag{7.5}$$

For the train (7.2)

$$\boldsymbol{g} = \boldsymbol{w}/\gamma \ . \tag{7.6}$$

Suppose now that $\boldsymbol{k}$ is replaced by $\boldsymbol{k} + i\epsilon\boldsymbol{k}'$ where ϵ is infinitesimal and $\boldsymbol{k}'$ is arbitrary and real. In order that (7.1) be satisfied, ω must be replaced by $\omega + i\epsilon\omega'$, where

$$\omega' = \boldsymbol{k}' \cdot \frac{\partial f}{\partial \boldsymbol{k}} = k_j' \frac{\partial f}{\partial k_j} \ . \tag{7.7}$$

Here the Taylor expansion of $f(\boldsymbol{k} + i\epsilon\boldsymbol{k}')$ has been used and terms higher than first order in ϵ have been neglected.

The wave train is now of the form

$$(\boldsymbol{C} + \epsilon\boldsymbol{C}') \exp i(\boldsymbol{k} \cdot \boldsymbol{x} - \omega t) \exp(-\epsilon\boldsymbol{k}' \cdot \boldsymbol{x} + \epsilon\omega' t) \ . \tag{7.8}$$

The corresponding energy flux is a function of ϵ, say $\boldsymbol{r}(\epsilon)$. Using assumption (1) we infer that it has the form

$$\begin{aligned} \boldsymbol{r}(\epsilon) = {} & [(\boldsymbol{A} + \epsilon\boldsymbol{\Phi}) \exp 2i(\boldsymbol{k} \cdot \boldsymbol{x} - \omega t) + (\overline{\boldsymbol{A}} + \epsilon\overline{\boldsymbol{\Phi}}) \exp -2i(\boldsymbol{k} \cdot \boldsymbol{x} - \omega t) \\ & + (\boldsymbol{w} + \epsilon\boldsymbol{h})] \exp[-2\epsilon\boldsymbol{k}' \cdot \boldsymbol{x} + 2\epsilon\omega' t] \ , \end{aligned} \tag{7.9}$$

where Φ is some bivector and $\boldsymbol{h}$ is some real vector. Similarly, e is also a function of ϵ, say $e(\epsilon)$, and has the form

$$e(\epsilon) = [(A + \epsilon B) \exp 2i(\boldsymbol{k} \cdot \boldsymbol{x} - \omega t) + (\overline{A} + \epsilon\overline{B}) \exp -2i(\boldsymbol{k} \cdot \boldsymbol{x} - \omega t) + (\gamma + \epsilon\delta)] \exp[-2\epsilon\boldsymbol{k}' \cdot \boldsymbol{x} + 2\epsilon\omega' t] , \tag{7.10}$$

where B is some complex scalar and δ some real scalar.

On inserting (7.9) and (7.10) into (6.6), the equation of conservation of energy, we find

$$(\boldsymbol{w} + \epsilon\boldsymbol{h}) \cdot \boldsymbol{k}' = (\gamma + \epsilon\delta)\omega' . \tag{7.11}$$

Hence

$$\boldsymbol{w} \cdot \boldsymbol{k}' = \gamma\omega' , \tag{7.12}$$

and using (7.7) this gives

$$(\boldsymbol{w} - \gamma\frac{\partial f}{\partial \boldsymbol{k}}) \cdot \boldsymbol{k}' = 0 . \tag{7.13}$$

This holds for arbitrary $\boldsymbol{k}'$ because no restrictions have been placed on $\boldsymbol{k}'$. Hence

$$\boldsymbol{w} = \gamma\frac{\partial f}{\partial \boldsymbol{k}} . \tag{7.14}$$

Thus, by (7.6)

$$\frac{\partial f}{\partial \boldsymbol{k}} = \frac{\boldsymbol{w}}{\gamma} = \boldsymbol{g} . \tag{7.15}$$

Thus for the train of elliptically polarised plane waves, the energy flux velocity is equal to $\dfrac{\partial\omega(\boldsymbol{k})}{\partial \boldsymbol{k}}$, which is called the **group velocity.**

For the three waves propagating in a general homogeneous anisotropic elastic body with frequency ω and wave-vector $\boldsymbol{k}$, it has been seen (§5) that $\omega = V_\alpha(\boldsymbol{k})$. The corresponding energy flux velocities are denoted by $\boldsymbol{g}^{(\alpha)}$; and from (7.15),

$$g_i^{(\alpha)} = \frac{\partial V_\alpha(\boldsymbol{k})}{\partial k_i} = \frac{\partial V_\alpha(\boldsymbol{s})}{\partial s_i} , \quad \alpha = 1, 2, 3 \tag{7.16}$$

on using the fact that V_α is homogeneous of degree one and thus its derivatives are homogeneous of degree zero.

Now $\partial V_\alpha(\boldsymbol{s})/\partial \boldsymbol{s}$ is normal to the sheet of the slowness surface $V_\alpha(\boldsymbol{s}) = 1$, because $\partial V_\alpha(\boldsymbol{s})/\partial s_i$ are the components of the gradient of (5.12) in the space of coordinates s_i. Thus it follows that the group velocity of a wave train propagating along $\boldsymbol{n}$ with phase speed $c = V_\alpha(\boldsymbol{n})$ is along the normal to the slowness surface $V_\alpha(\boldsymbol{s}) = 1$ at the "point" $\boldsymbol{s} = (1/V_\alpha)\boldsymbol{n}$.

For isotropic media, the slowness surface consists of the two concentric spheres $\rho = (\lambda + 2\mu)s^2, \rho = \mu s^2$. The corresponding energy flux velocities are accordingly along $\boldsymbol{n}$, the direction of propagation.

Now we consider briefly linearly polarised waves in crystals and compute directly the mean energy flux vector and the normal to the slowness surface.

With $\boldsymbol{A} = \boldsymbol{a}$, where $\boldsymbol{a}$ is a real vector, the propagation condition (5.1) may also be written

$$(q_{ik}(\boldsymbol{k}) - \rho\omega^2\delta_{ik})a_k = 0 \,. \tag{7.17}$$

Taking the derivative with respect to k_p yields

$$\left(\frac{\partial q_{ij}}{\partial k_p} - 2\rho\omega\frac{\partial \omega}{\partial k_p}\delta_{ij}\right)a_j + (q_{ij} - \rho\omega^2\delta_{ij})\frac{\partial a_j}{\partial k_p} = 0 \,. \tag{7.18}$$

Hence

$$\frac{\partial \omega}{\partial k_p} = \frac{1}{2\rho\omega}\widehat{a}_i\frac{\partial q_{ij}}{\partial k_p}\widehat{a}_j \,, \tag{7.19}$$

where $\widehat{\boldsymbol{a}}$ is the unit vector along $\boldsymbol{a}$. But

$$\frac{\partial q_{ij}}{\partial k_p} = (c_{pijt} + c_{pjit})k_t \,, \tag{7.20}$$

and so the normal to the slowness surface is along

$$\frac{\partial \omega}{\partial k_p} = \frac{1}{\rho\omega}c_{pijt}\widehat{a}_i\widehat{a}_j k_t = \frac{1}{\rho v}c_{pijt}\widehat{a}_i\widehat{a}_j n_t \,. \tag{7.21}$$

Now using (6.9) and (6.10), with $\boldsymbol{A} = \boldsymbol{a}$, we have

$$\widetilde{r}_p/\widetilde{e} = (\rho v)^{-1}c_{pijt}\widehat{a}_i\widehat{a}_j n_t \,. \tag{7.22}$$

Thus $\boldsymbol{g} = \widetilde{\boldsymbol{r}}/\widetilde{e}$, confirming the general result in this special case.

Remark For propagation along an arbitrary direction $\boldsymbol{n}$ with a common frequency ω, if c_α are the phase velocities and $\boldsymbol{a}^{(\alpha)}$ are the mutually orthogonal real eigenvectors of the acoustical tensor, and if $\boldsymbol{g}^{(\alpha)} = \widetilde{\boldsymbol{r}}^{(\alpha)}/\widetilde{e}^{(\alpha)}$ are the energy flux velocities, or group velocities, corresponding to the three waves, then (Boulanger & Hayes 1993)

$$c_1 g_i^{(1)} + c_2 g_i^{(2)} + c_3 g_i^{(3)} = (1/\rho)c_{ijjl}n_l \,. \tag{7.23}$$

This is because

$$2\widetilde{r}_i^{(\alpha)} = (\omega^2/c_\alpha)c_{ijkl}a_j^{(\alpha)}a_k^{(\alpha)}n_l \,, \quad 2\widetilde{e}^{(\alpha)} = \rho\omega^2\boldsymbol{a}^{(\alpha)}\cdot\boldsymbol{a}^{(\alpha)} \,, \tag{7.24}$$

and hence

$$c_\alpha \widetilde{g}_i^{(\alpha)} = (1/\rho)c_{ijkl}\widehat{a}_j^{(\alpha)}\widehat{a}_k^{(\alpha)}n_l \,, \tag{7.25}$$

where $\widehat{\boldsymbol{a}}^{(\alpha)}$ is the unit vector along $\boldsymbol{a}^{(\alpha)}$. Then condition (7.23) follows because $\sum\limits_\alpha \widehat{a}_j^{(\alpha)}\widehat{a}_k^{(\alpha)} = \delta_{jk}$.

8. Examples

We consider some examples.

8.1 Waves in transversely isotropic materials

The acoustical tensor (5.3) for transversely-isotropic materials may be obtained by inserting (4.1) into the equations of motion (7.2). We find

$$\begin{aligned} q_{11}(\boldsymbol{n}) &= d_{11}n_1^2 + \{(d_{11} - d_{12})/2\}n_2^2 + d_{44}n_3^2 , \\ q_{22}(\boldsymbol{n}) &= \{(d_{112} - d_{12})/2\}n_1^2 + d_{11}n_2^2 + d_{44}n_3^2 , \\ 2q_{12}(\boldsymbol{n}) &= (d_{11} + d_{12})n_1n_2 , \\ q_{13}(\boldsymbol{n}) &= (d_{13} + d_{44})n_1n_3, \quad q_{23}(\boldsymbol{n}) = (d_{13} + d_{44})n_2n_3 , \\ q_{33}(\boldsymbol{n}) &= d_{44}(n_1^2 + n_2^2) + d_{33}n_3^2 . \end{aligned} \tag{8.1}$$

One root of the secular equation is

$$\rho c^2 = \{(d_{11} - d_{12})/2\}(n_1^2 + n_2^2) + d_{44}n_3^2 , \tag{8.2}$$

and the corresponding amplitude eigenvector of $q_{ik}(\boldsymbol{n})$, is, up to a scalar factor,

$$\boldsymbol{a} = \boldsymbol{n} \wedge \boldsymbol{k} = (n_2, -n_1, 0) , \tag{8.3}$$

where $\boldsymbol{k}$ is the unit vector along the symmetry axis Oz. That the secular equation has this simple factor was noticed by Synge (1957) following the publication of slowness diagrams by Musgrave (1954) for various crystals. Musgrave's diagrams showed for the examples considered that there was a separable sheet of the slowness surface.

The two other roots $\rho c_2^2, \rho c_3^2$ of the secular equation are the solutions of the quadratic equation

$$\begin{aligned} &(\rho c^2)^2 - \{(d_{11} + d_{44})(n_1^2 + n_2^2) + (d_{33} + d_{44})n_3^2\}\rho c^2 \\ &\quad + d_{11}d_{44}(n_1^2 + n_2^2)^2 + \{d_{11}d_{33} + d_{44}^2 - (d_{13} + d_{44})^2\}(n_1^2 + n_2^2)n_3^2 \\ &\quad + d_{33}d_{44}n_3^4 = 0 . \end{aligned} \tag{8.4}$$

Thus the slowness surface consists of an ellipsoid (corresponding to (8.2)) and of a double surface symmetrical about Oz.

In order to study the propagation in a direction orthogonal to the symmetry axis we may take $\boldsymbol{n}$ along Ox without loss in generality. Then the solutions are the longitudinal wave

$$\boldsymbol{u} = \boldsymbol{i}\{\alpha \exp ik(x - \sqrt{d_{11}/\rho}t)\}^+ , \tag{8.5}$$

and the two transverse waves

$$\begin{aligned} \boldsymbol{u} &= \boldsymbol{j}\{\beta \exp ik(x - \sqrt{(d_{11} - d_{12})/2\rho t})\}^+ , \\ \boldsymbol{u} &= \boldsymbol{k}\{\gamma \exp ik(x - \sqrt{(d_{44}/\rho)t}\}^+ , \end{aligned} \tag{8.6}$$

where α, β, γ are arbitrary constants. Because the slowness surface is symmetrical about Oz it follows that the normals to it at $\mathrm{s} = s\boldsymbol{i}$, are along $\boldsymbol{i}$, and hence the mean energy flux vectors corresponding to these three waves is along Ox, the direction of propagation.

For propagation along Oz, the solutions are the longitudinal wave

$$\boldsymbol{u} = \boldsymbol{k}\{\gamma \exp ik(z - \sqrt{d_{33}/\rho t})\}^+ , \tag{8.7}$$

and the transverse wave

$$\boldsymbol{u} = \{(\alpha\boldsymbol{i} + \beta\boldsymbol{j}) \exp ik(z - \sqrt{d_{44}/\rho t})\}^+ , \tag{8.8}$$

corresponding to the double root of the secular equation, $\rho c^2 = d_{44}$. Again α, β and γ are arbitrary constants. In general the wave (8.8) is elliptically polarized. It is circularly polarized if $\beta/\alpha = \pm i$.

Using (3.3), it follows for the solution (8.7) that $v_1 = v_2 = 0$, $t_{13} = t_{23} = 0$, and hence $r_1 = r_2 = 0$. Thus the energy flux vector is along Oz. Similarly for (8.8), on using (3.3), we have $v_3 = 0, t_{11} = t_{22} = t_{12} = 0$ and hence $r_1 = r_2 = 0$, so that $\boldsymbol{r}$ is along Oz.

For propagation in a general direction the waves are neither purely longitudinal nor purely transverse.

Remark 1. In general the secular equation may be factored for all $\boldsymbol{n}$ only when the material is isotropic or transversely isotropic.

Remark 2. Boulanger & Hayes (1993) have examined the conditions under which circularly polarised waves may propagate in transversely isotropic materials.

8.2 Waves in cubic crystals

For propagation along Ox, we find from (3.6), the longitudinal wave solution

$$\boldsymbol{u} = \boldsymbol{i}\{\alpha \exp ik(x - \sqrt{c_{11}/\rho t})\}^+ , \tag{8.9}$$

and the transverse wave

$$\boldsymbol{u} = \{(\beta\boldsymbol{j} + \gamma\boldsymbol{k}) \exp ik(x - \sqrt{c_{44}/\rho t})\}^+ , \tag{8.10}$$

where α, β, γ are arbitrary constants. In general (8.10) is elliptically polarised. It is circularly polarised if $\beta/\gamma = \pm i$.

For the linearly polarised wave (8.9) we have $v_2 = v_3 = 0$, and from (3.5) $t_{21} = t_{31} = 0$ so that $r_2 = r_3 = 0$ and accordingly $\boldsymbol{r}$ is along Ox. Similarly, for (8.10), $v_1 = 0$ and $t_{22} = t_{33} = t_{23} = 0$, so that again $\boldsymbol{r}$ is along Ox.

Similarly, a longitudinal wave with phase speed $\sqrt{c_{11}/\rho}$ may propagate along Oy and Oz , and an elliptically polarized wave with phase speed $\sqrt{c_{44}/\rho}$ may propagate along Oy and Oz. In every case the energy flux vector is along the direction of propagation.

For propagation along $\boldsymbol{n}^* = (1,1,1)/\sqrt{3}$ it is easily seen that the secular equation has the simple root

$$\rho c^2 = (c_{11} + 2c_{12} + 4c_{44})/3 \ , \quad \boldsymbol{A} \| \boldsymbol{n}^* \ , \tag{8.11}$$

and the double root

$$\rho c^2 = (c_{11} - c_{12} + c_{44})/3 \ , \quad \boldsymbol{A} \cdot \boldsymbol{n}^* = 0 \ . \tag{8.12}$$

Thus propagation along $\boldsymbol{n}^*$ is similar to that in an isotropic body - a purely longitudinal wave and an elliptically polarised wave may propagate.

In particular, linearly polarised waves with any displacement vector transverse to the wave normal $\boldsymbol{n}^*$ may be propagated with the speed given by $(8.12)_1$. As the direction of the displacement vector is varied in the plane whose normal is $\boldsymbol{n}^*$ the corresponding mean-energy flux vector deviates from $\boldsymbol{n}^*$ and trace out a circular cone about $\boldsymbol{n}^*$. Musgrave (1970) shows that the semi-angle of this cone is $\tan^{-1}\{(c_{11}-c_{12}-2c_{44})/[(c_{11}-c_{12}+c_{44})\sqrt{2}]\}$. The associated physical phenonemon is called "internal conical refraction" and has been observed experimentally (de Klerk & Musgrave 1955).

8.3 Waves in inextensible materials

We seek solutions

$$T = \{B \exp ik(\boldsymbol{n} \cdot \boldsymbol{x} - ct)\}^+ \ , \ \boldsymbol{u} = \{\boldsymbol{A} \exp ik(\boldsymbol{n} \cdot \boldsymbol{x} - ct)\}^+ \ . \tag{8.13}$$

The equations of motion (3.9) and the inextensibility condition $(3.8)_2$ lead to

$$\begin{aligned} ik^{-1}B(\boldsymbol{l} \cdot \boldsymbol{n})l_i + q_{ik}(\boldsymbol{n})A_k &= \rho c^2 A_i \ , \\ (\boldsymbol{l} \cdot \boldsymbol{n})(\boldsymbol{l} \cdot \boldsymbol{A}) &= 0 \ , \end{aligned} \tag{8.14}$$

where $\boldsymbol{q}(\boldsymbol{n})$, which is symmetric, is now given by

$$q_{ik}(\boldsymbol{n}) = d_{ijkl} n_j n_l \ . \tag{8.15}$$

We note that if $\boldsymbol{n}$, the propagation direction, is orthogonal to $\boldsymbol{l}$, the direction of inextensibility, so that $\boldsymbol{l}\cdot\boldsymbol{n}=0$, then $(8.14)_2$ is satisfied and $(8.14)_1$ reduces to the propagation condition of an unconstrained material. Thus, corresponding to a propagation direction $\boldsymbol{n}$, orthogonal to $\boldsymbol{l}$, there are in general three linearly polarised homogeneous waves, the amplitude vectors being mutually orthogonal.

However, for any direction $\boldsymbol{n}$ which is not orthogonal to $\boldsymbol{l}$, $(8.14)_2$ gives $\boldsymbol{l}\cdot\boldsymbol{A}=0$. Then, from (8.14),

$$ik^{-1}B(\boldsymbol{l}\cdot\boldsymbol{n}) = -q_{ik}(\boldsymbol{n})l_i A_k \ , \tag{8.16}$$

so that

$$(\delta_{ir} - l_i l_r)q_{rs}A_s = \rho c^2 A_i \ , \tag{8.17}$$

or, alternatively,

$$q'_{is}(\boldsymbol{n})A_s = \rho c^2 A_i \ , \tag{8.18}$$

where

$$q'_{is}(\boldsymbol{n}) = (\delta_{ir} - l_i l_r)q_{rm}(\delta_{ms} - l_m l_s) \ . \tag{8.19}$$

Clearly $\boldsymbol{l}$ is an eigenvector of $\boldsymbol{q}'$ corresponding to a zero eigenvalue. We note that $\boldsymbol{q}'$ is real and symmetric. Thus the two other eigenvalues ρc_1^2 and ρc_2^2 are both real and assuming $c_1^2 \neq c_2^2$ the corresponding amplitudes are orthogonal to each other and to $\boldsymbol{l}$. From this we conclude that, in general, in every direction $\boldsymbol{n}$ which is not orthogonal to $\boldsymbol{l}$, two linearly polarised waves may propagate with amplitude vectors orthogonal to each other and to $\boldsymbol{l}$. subsectionWaves in Incompressible materials

We seek solutions of the form

$$p = \{P \exp ik(\boldsymbol{n}\cdot\boldsymbol{k} - ct)\}^+ \ , \quad \boldsymbol{u} = \{\boldsymbol{A}\exp ik(\boldsymbol{n}\cdot\boldsymbol{x} - ct)\}^+ \ . \tag{8.20}$$

The equations of motion (3.11) and the constraint $(3.10)_2$, lead to

$$ik^{-1}Pn_i + q_{ik}(\boldsymbol{n})A_k = \rho c^2 A_i \ , \quad A_i n_i = 0 \ , \tag{8.21}$$

where $\boldsymbol{q}$ has the form (8.15). Using $(8.21)_2$ we find

$$P = ikn_i q_{ik} A_k \ , \tag{8.22}$$

and hence

$$(\delta_{ip} - n_i n_p)q_{pk}A_k = \rho c^2 A_i \ , \tag{8.23}$$

or alternatively

$$q''_{ik}A_k = \rho c^2 A_i \ , \tag{8.24}$$

where

$$q''_{ik} = (\delta_{ip} - n_i n_p)q_{pr}(\delta_{rk} - n_r n_k) \ . \tag{8.25}$$

Clearly $\boldsymbol{n}$ is an eigenvector of $\boldsymbol{q}''$ corresponding to a zero eigenvalue. We note that $\boldsymbol{q}''$ is real and symmetric. Thus the two other eigenvalues $\rho c_1^2, \rho c_2^2$ are both real and assuming $c_1^2 \neq c_2^2$ the corresponding amplitudes are orthogonal to each other and to $\boldsymbol{n}$. Thus in general two purely transverse linearly polarised waves may propagate in every direction $\boldsymbol{n}$ in a homogeneous anisotropic incompressible elastic material.

9. Inhomogeneous Plane Waves

For inhomogeneous plane waves the elastic displacement is given by the real part of the expression $\{\boldsymbol{A} \exp i\omega(\boldsymbol{S} \cdot \boldsymbol{x} - t)\}$. Here ω is the real angular frequency of the waves, and $\boldsymbol{A} = \boldsymbol{A}^+ + i\boldsymbol{A}^-$ is a complex vector or bivector to use the term of Hamilton (1853) and Gibbs (1881). We call $\boldsymbol{A}$ the 'amplitude bivector'. Also $\boldsymbol{S} = \boldsymbol{S}^+ + i\boldsymbol{S}^-$ is a bivector, the 'slowness bivector'. The planes of constant phase are $\boldsymbol{S}^+ \cdot \boldsymbol{x} =$ constant and the planes of constant amplitude are $\boldsymbol{S}^- \cdot \boldsymbol{x} =$ constant. The phase slowness is $|\boldsymbol{S}^+|$ and the attenuation factor is $|\boldsymbol{S}^-|$. Thus inhomogeneous plane waves are described in terms of two bivectors. It is because of this that the study of bivectors is intimately connected with the study of inhomogeneous plane waves.

Hamilton makes passing reference to bivectors in his book on quaternions. His contemporary, MacCullagh (1847), in his work on electromagnetic waves derived a central result on bivectors - though not in the context of bivectors. However, Gibbs was the first to make a systematic study of bivectors. He published privately a seventy three page pamphlet "Elements of Vector Analysis" of which seven pages are devoted to bivectors. Gibbs associated an ellipse with each bivector: if $\boldsymbol{A} = \boldsymbol{c} + i\boldsymbol{d}$ then the ellipse associated with $\boldsymbol{A}$ has the real vectors $\boldsymbol{c}$ and $\boldsymbol{d}$ as a pair of conjugate radii. Gibbs gave a geometrical interpretation to the scalar product of two bivectors being zero. Synge (1964) gave a systematic treatment of the algebra of bivectors, but apparently unaware of Gibbs' work did not give geometrical significance to his results. Stone (1963) makes use of bivectors in describing propagation of waves. Indeed, the ellipse associated with the amplitude bivector of a plane wave is the polarisation ellipse of this wave.

10. The Ellipse

A pair of diameters in an ellipse is said to be **conjugate** if all chords parallel to one diameter are bisected by the other diameter. Equivalently the tangent at the extremity of one diameter is parallel to the other diameter.

Let OM and ON be a pair of conjugate radii of an ellipse, centre O. Taking (oblique) axes Ox and Oy along OM and ON, respectively, the equation of the ellipse is $x^2/a^2 + y^2/b^2 = 1$, where $OM = a$, $ON = b$. A typical point P on the ellipse has coordinates $x = a\cos\phi$, $y = b\sin\phi$, so that the position vector of P is

$$\boldsymbol{r} = \boldsymbol{OP} = a\cos\phi \boldsymbol{i}^* + b\sin\phi \boldsymbol{j}^* ,$$

where $\boldsymbol{i}^*$ and $\boldsymbol{j}^*$ are unit vectors along OM and ON respectively. This equation may also be written

$$\boldsymbol{r} = \boldsymbol{a}\cos\phi + \boldsymbol{b}\sin\phi ,$$

where $\boldsymbol{a} = \boldsymbol{OM} = a\boldsymbol{i}^*$, $\boldsymbol{b} = \boldsymbol{ON} = b\boldsymbol{j}^*$. We note that $\boldsymbol{r}(0) = \boldsymbol{a}$, $\boldsymbol{r}(\pi/2) = \boldsymbol{b}$, and the tangent is along $(d\boldsymbol{r}/d\phi) = -\boldsymbol{a}\sin\phi + \boldsymbol{b}\cos\phi$. Thus $\boldsymbol{r}(\phi + \pi/2)$ is parallel to the tangent at ϕ, and so $\boldsymbol{r}(\phi)$ and $\boldsymbol{r}(\phi + \pi/2)$ are conjugate radii.

11. Bivectors

If $\boldsymbol{a}$ and $\boldsymbol{b}$ are two real vectors, then $\boldsymbol{C} = \boldsymbol{a} + i\boldsymbol{b}$ is called a bivector. The two bivectors $\boldsymbol{C}$, and $\boldsymbol{D} = \boldsymbol{p} + i\boldsymbol{q}$, are equal if and only if $\boldsymbol{a} = \boldsymbol{p}$ and $\boldsymbol{b} = \boldsymbol{q}$.

The dot product of the two bivectors $\boldsymbol{C}$ and $\boldsymbol{D}$ is defined as follows:

$$\boldsymbol{C} \cdot \boldsymbol{D} = (\boldsymbol{a} + i\boldsymbol{b}) \cdot (\boldsymbol{p} + i\boldsymbol{q}) = \boldsymbol{a} \cdot \boldsymbol{p} - \boldsymbol{b} \cdot \boldsymbol{q} + i(\boldsymbol{b} \cdot \boldsymbol{p} + \boldsymbol{a} \cdot \boldsymbol{q}) .$$

Similarly the cross product is defined:

$$\boldsymbol{C} \wedge \boldsymbol{D} = (\boldsymbol{a} + i\boldsymbol{b}) \wedge (\boldsymbol{p} + i\boldsymbol{q}) = \boldsymbol{a} \wedge \boldsymbol{p} - \boldsymbol{b} \wedge \boldsymbol{q} + i(\boldsymbol{a} \wedge \boldsymbol{q} + \boldsymbol{b} \wedge \boldsymbol{p}) .$$

Any pair of vectors may be considered as conjugate radii of an ellipse. The ellipse is uniquely determined if its centre is at the common origin of the two vectors. Thus the ellipse of $\boldsymbol{C}$ has $\boldsymbol{a}$ and $\boldsymbol{b}$ as a pair of conjugate radii. The sense of description is from the terminus of $\boldsymbol{C}^+(= \boldsymbol{a})$ to the terminus of $\boldsymbol{C}^-(= \boldsymbol{b})$. The ellipse of $\overline{\boldsymbol{C}} \equiv \boldsymbol{a} - i\boldsymbol{b}$, the complex conjugate of $\boldsymbol{C}$, is the same as the ellipse of $\boldsymbol{C}$ but is described from the terminus of $\overline{\boldsymbol{C}}^+(= \boldsymbol{a})$ towards the terminus of $\overline{\boldsymbol{C}}^-(= -\boldsymbol{b})$ i.e. in the opposite sense to that of the ellipse of $\boldsymbol{C}$.

MacCullagh's Theorem (1842)

If $(\boldsymbol{a}, \boldsymbol{b})$ is a pair of conjugate radii of the ellipse of $\boldsymbol{C}$, that is points on the ellipse of $\boldsymbol{C}$ are given by

$$\boldsymbol{r} = \boldsymbol{a}\cos\theta + \boldsymbol{b}\sin\theta , \qquad 0 \le \theta \le 2\pi , \tag{11.1}$$

then $(\boldsymbol{h}, \boldsymbol{l})$, given by $\boldsymbol{h} + i\boldsymbol{l} = e^{i\phi}\boldsymbol{C} = (\cos\phi + i\sin\phi)(\boldsymbol{a} + i\boldsymbol{b})$, is also a pair of conjugate radii of the ellipse of $\boldsymbol{C}$.

Proof. The equation (11.1) may be written

$$\boldsymbol{r} = \boldsymbol{h}\cos(\theta + \phi) + \boldsymbol{l}\sin(\theta + \phi) , \qquad 0 \le \theta + \phi \le 2\pi ,$$

where

$$\boldsymbol{h} = \cos\phi\ \boldsymbol{a} - \sin\phi\ \boldsymbol{b} , \quad \boldsymbol{l} = \sin\phi\ \boldsymbol{a} + \cos\phi\ \boldsymbol{b}. \tag{11.2}$$

Hence $(\boldsymbol{h}, \boldsymbol{l})$ is also a pair of conjugate radii of the ellipse (11.1).

Thus, in particular, any bivector $\boldsymbol{A}$ may be written in the form

$$\boldsymbol{A} = e^{i\phi}(\boldsymbol{p} + i\boldsymbol{q}) , \qquad \boldsymbol{p} \cdot \boldsymbol{q} = 0 ,$$

where $\boldsymbol{p}$ and $\boldsymbol{q}$ are along the principal axes of the ellipse of $\boldsymbol{A}$.

12. Multiplication by a Scalar

Let $\boldsymbol{A} = \boldsymbol{c} + i\boldsymbol{d}$, $\boldsymbol{B} = \boldsymbol{p} + i\boldsymbol{q}$. Then $\boldsymbol{A} \pm \boldsymbol{B} = (\boldsymbol{c} \pm \boldsymbol{p}) + i(\boldsymbol{d} \pm \boldsymbol{q})$ and the ellipse of $\boldsymbol{A} \pm \boldsymbol{B}$ is

$$\boldsymbol{r} = (\boldsymbol{c} \pm \boldsymbol{p}) \cos\theta + (\boldsymbol{d} \pm \boldsymbol{q}) \sin\theta .$$

For n bivectors $\boldsymbol{A}_j = \boldsymbol{c}_j + i\boldsymbol{d}_j$., their sum is $\sum \boldsymbol{A}_j = \sum \boldsymbol{c}_j + i \sum \boldsymbol{d}_j$ and the corresponding ellipse is

$$\boldsymbol{r} = (\sum \boldsymbol{c}_j) \cos\theta + (\sum \boldsymbol{d}_j) \sin\theta .$$

This means that for any number of (inhomogeneous) waves all propagating with the same frequency, the polarisation ellipse of the resultant field at any point $\boldsymbol{x}$ can be written down immediately.

If $\boldsymbol{C} = \boldsymbol{a} + i\boldsymbol{b}$, and $\alpha = |\alpha| e^{i\phi}$, is an arbitrary scalar, then

$$\alpha \boldsymbol{C} = |\alpha| e^{i\phi}(\boldsymbol{a} + i\boldsymbol{b}) = |\alpha|(\boldsymbol{h} + i\boldsymbol{l})$$

where $(\boldsymbol{h}, \boldsymbol{l})$ is a pair of conjugate radii of the ellipse of $\boldsymbol{C}$. Hence the effect of multiplication of a bivector by a complex scalar α is to rotate (non-uniformly) the pair $(\boldsymbol{a}, \boldsymbol{b})$ of conjugate radii of the ellipse of $\boldsymbol{C}$ into another pair of conjugate radii of the ellipse of $\boldsymbol{C}$, and then each member of this pair is subjected to a uniform extension $|\alpha|$.

Two bivectors $\boldsymbol{A}$ and $\boldsymbol{B}$ are said to be **parallel** if there exists a scalar α such that $\boldsymbol{A} = \alpha \boldsymbol{B}$. In this case the major and minor axes of the ellipse of $\boldsymbol{A}$ are parallel respectively to the major and minor axes of the ellipse of $\boldsymbol{B}$ and also the aspect ratio (i.e. the ratio of the length of the major axis to the length of the minor axis) is the same for each ellipse. Further, the ellipses are described in the same sense. We say that the directional ellipses of $\boldsymbol{A}$ and $\boldsymbol{B}$ are similar and similarly situated.

If two bivectors are not parallel they are said to be **linearly independent**.

13. The Dot Product

Two bivectors are said to be orthogonal if their dot product is zero. If $\boldsymbol{A} \cdot \boldsymbol{B} = 0$, then either $\boldsymbol{A} = 0$, or $\boldsymbol{B} = 0$, or the projection of the ellipse of $\boldsymbol{A}$ upon the plane of the ellipse of $\boldsymbol{B}$ is an ellipse which is similar and similarly situated to the ellipse of $\boldsymbol{B}$ when rotated through a quadrant.

Indeed, suppose $\boldsymbol{A} = e^{i\phi}(\boldsymbol{a} + i\boldsymbol{b})$, where $\boldsymbol{a} \cdot \boldsymbol{b} = 0$, $|\boldsymbol{a}| > |\boldsymbol{b}|$. Any other bivector $\boldsymbol{B}$ may then be written $\boldsymbol{B} = p\boldsymbol{a} + q\boldsymbol{b} + r(\boldsymbol{a} \wedge \boldsymbol{b})$ for some complex scalars p, q, r. Then $\boldsymbol{A} \cdot \boldsymbol{B} = 0$ implies that $p\boldsymbol{a} \cdot \boldsymbol{a} + iq\boldsymbol{b} \cdot \boldsymbol{b} = 0$ and hence

$$\boldsymbol{B} = \{-iq((\boldsymbol{b} \cdot \boldsymbol{b})\boldsymbol{a} + i(\boldsymbol{a} \cdot \boldsymbol{a})\boldsymbol{b})/(\boldsymbol{a} \cdot \boldsymbol{a})\} + r(\boldsymbol{a} \wedge \boldsymbol{b}) .$$

The term inside the chain bracket is the orthogonal projection of $\boldsymbol{B}$ upon the plane of the ellipse of $\boldsymbol{A}$. The corresponding ellipse has major axis along $\boldsymbol{b}$, minor axis along $\boldsymbol{a}$, aspect ratio $|\boldsymbol{b}|/|\boldsymbol{a}|$, and is described in the same sense as the ellipse of $\boldsymbol{A}$.

14. Isotropic Bivectors

. A bivector $\boldsymbol{A}$ is said to be **isotropic** if $\boldsymbol{A} \cdot \boldsymbol{A} = 0$, the corresponding ellipse now being a circle. Typically, an isotropic bivector may be written $\boldsymbol{A} = \alpha(\boldsymbol{i} + i\boldsymbol{j})$, where α is a scalar. Every bivector $\boldsymbol{B}$ orthogonal to this has the form

$$\boldsymbol{B} = \beta(\boldsymbol{i} + i\boldsymbol{j}) + \gamma\boldsymbol{k} ,$$

where β and γ are arbitrary complex scalars. If this bivector $\boldsymbol{B}$ is also isotropic, then $\gamma = 0$ from which it is concluded that two non-parallel isotropic bivectors cannot be orthogonal; or equivalently if two isotropic bivectors are orthogonal they are parallel (Synge, 1964).

Also, it may be shown that if three non-zero linearly independent bivectors are mutually orthogonal none is an isotropic bivector (Synge, 1964).

15. Homogeneous and Inhomogeneous Plane Waves. Prescription.

The link between bivectors and inhomogeneous plane waves is that the field, for example, the displacement field $\boldsymbol{u}$, corresponding to a train of such waves is described in terms of two bivectors, the amplitude and slowness bivectors. Thus $\boldsymbol{u} = [\boldsymbol{A} \exp i\omega(\boldsymbol{S} \cdot \boldsymbol{x} - t)]^+$, where $\boldsymbol{A}$ is the amplitude, and $\boldsymbol{S}$ the slowness.

For non-attenuated homogeneous waves the field has the form $\boldsymbol{u} = [\boldsymbol{A} \exp i(\boldsymbol{k} \cdot \boldsymbol{x} - \omega t)]^+$ where $\boldsymbol{k} = k\boldsymbol{n}$ is the real wave-vector and the phase speed is $\omega/|\boldsymbol{k}|$. Typically, for homogeneous waves, the direction of $\boldsymbol{k}$ is specified and an eigenvalue problem is solved to determine the corresponding phase speeds and the corresponding amplitudes $\boldsymbol{A}$. For inhomogeneous waves, on the other hand, the directions of the normals to the planes of constant phase and the normals to the planes of constant amplitude, may not be chosen arbitrarily. For example, in an isotropic elastic medium the planes of constant phase must be orthogonal to the planes of constant amplitude whilst in an isotropic viscoelastic medium these planes may not be orthogonal. Instead, as suggested by Hayes (1984), we write $\boldsymbol{S} = N\boldsymbol{C}$, where $\boldsymbol{C} = m\widehat{\boldsymbol{m}} + i\widehat{\boldsymbol{n}}$ with $\widehat{\boldsymbol{m}} \cdot \widehat{\boldsymbol{n}} = 0$, $|\widehat{\boldsymbol{m}}| = |\widehat{\boldsymbol{n}}| = 1$, and m is a real scalar. If $\boldsymbol{C}$ is prescribed, then N and the corresponding amplitude bivector may be determined from an eigenvalue problem. Prescription of $\boldsymbol{C}$ is equivalent to the prescription of an ellipse - the directional ellipse, to use Gibbs' phrase - whose principal axes are of length $|m|$ and 1 , and lie along $\widehat{\boldsymbol{m}}$ and $\widehat{\boldsymbol{n}}$, respectively. Thus whereas a direction is prescribed in the case of homogeneous waves, a directional ellipse $\boldsymbol{C}$ is prescribed in the case of inhomogeneous waves. Let $N = Te^{i\phi}$ where T and ϕ are real. Then, for given $\boldsymbol{C}$, if N is known, the directions of the normals to the planes of constant phase and the normals to the planes of constant amplitude are determined. Indeed $\boldsymbol{S}^+ = T(e^{i\phi}\boldsymbol{C})^+$, $\boldsymbol{S}^- = T(e^{i\phi}\boldsymbol{C})^-$, so that the normals to the planes of constant phase and the normals to the planes of constant amplitude lie along a pair of conjugate directions of the directional ellipse.

To find all homogeneous plane wave solutions the direction of $\boldsymbol{k}$ is altered to take in every possible direction. Similarly, to find all possible inhomogeneous plane wave solutions, the directional ellipse $\boldsymbol{C}$ is altered to take in every possibility. First of all, $\hat{\boldsymbol{n}}$ is chosen, then $\widehat{\boldsymbol{m}}$ takes any direction in the plane orthogonal to $\hat{\boldsymbol{n}}$, whilst m takes on all possible real values. Then $\hat{\boldsymbol{n}}$ is altered and the process repeated. As $\hat{\boldsymbol{n}}$ takes on every possible direction and $\widehat{\boldsymbol{m}}$ is varied in direction and m varied in magnitude, all possible inhomogeneous plane wave solutions are obtained.

16. Inhomogeneous Waves in Crystals

We assume

$$\boldsymbol{u} = [\boldsymbol{A} \exp i\omega(\boldsymbol{S}\cdot\boldsymbol{x} - t)]^+ , \quad \boldsymbol{S} = N\boldsymbol{C}, \quad \boldsymbol{C} = m\widehat{\boldsymbol{m}} + i\hat{\boldsymbol{n}}, \quad N = Te^{i\phi} , \tag{16.1}$$

where $\widehat{\boldsymbol{m}}\cdot\hat{\boldsymbol{n}} = 0$, and ω is real. Insertion of (16.1) into (2.6) leads to the propagation condition

$$Q_{ik}(\boldsymbol{C})A_k = \rho N^{-2} A_i , \tag{16.2}$$

where

$$Q_{ik}(\boldsymbol{C}) = c_{ijkl}C_jC_l . \tag{16.3}$$

This is an eigenvalue problem for the complex symmetric tensor $\boldsymbol{Q}(\boldsymbol{C})$. The secular equation is

$$\det(\boldsymbol{Q}(\boldsymbol{C}) - \rho N^{-2}\mathbf{1}) = 0 , \tag{16.4}$$

a cubic in N^{-2} for given $\boldsymbol{C}$.

For given $\boldsymbol{C}$, let the roots of (16.4) be denoted by $N_\alpha^{-2}, \alpha = 1,2,3$, and let the corresponding eigenbivectors be denoted by $\boldsymbol{A}^{(\alpha)}$. On using the symmetry of $\boldsymbol{Q}$, it follows from (16.2), that if $N_\alpha^{-2} \neq N_\beta^{-2}$, then

$$\boldsymbol{A}^{(\alpha)} \cdot \boldsymbol{A}^{(\beta)} = 0, \quad \alpha \neq \beta; \ \alpha,\beta = 1,2,3 . \tag{16.5}$$

Thus, in general, the plane of the ellipse of $\boldsymbol{A}^{(\alpha)}$ may not be orthogonal to the plane of the ellipse of $\boldsymbol{A}^{(\beta)}$. If the three roots are all different, the ellipses of $\boldsymbol{A}^{(\alpha)}$ and $\boldsymbol{A}^{(\beta)}$ when projected upon the plane of the ellipse of $\boldsymbol{A}^{(\gamma)}$ are similar and similarly situated ellipses and described in the same sense; and similar and similarly situated also to the ellipse of $\boldsymbol{A}^{(\gamma)}$ when rotated through a quadrant.

Propagation in internally constrained media can be developed in the same way as homogeneous wave propagation was examined for inextensible materials in §8.3 and for incompressible materials in §8.4.

17. Isotropic Eigenbivectors

It has been seen that the determination of the N and the corresponding $\boldsymbol{A}$ for given $\boldsymbol{C}$ reduces to an eigenvalue problem. Here we recall the proof that if $\boldsymbol{Q}(\boldsymbol{C})$ has an isotropic eigenbivector then the corresponding eigenvalue must be double. The reader is referred to the original paper (Hayes, 1984) for a proof of the converse.

Given that $\boldsymbol{Q} = \boldsymbol{Q}^T$ and that

$$\boldsymbol{QA} = \lambda \boldsymbol{A}, \quad \boldsymbol{A} \cdot \boldsymbol{A} = 0 \,, \tag{17.1}$$

we are to show λ must be a double root of $\det(\boldsymbol{Q} - \lambda \mathbf{1}) = 0$.

Write

$$\boldsymbol{A} = \boldsymbol{r} + i\boldsymbol{s} \,, \quad \boldsymbol{r} \cdot \boldsymbol{r} = \boldsymbol{s} \cdot \boldsymbol{s} = 1, \quad \boldsymbol{r} \cdot \boldsymbol{s} = 0 \,, \tag{17.2}$$

without loss in generality. Let $\boldsymbol{t}$ be chosen to complete an orthonormal triad with $\boldsymbol{r}$ and $\boldsymbol{s}$. Referred to this triad let $\boldsymbol{Q}$ have the form $\widehat{\boldsymbol{Q}}$, given by

$$\widehat{\boldsymbol{Q}} = \begin{bmatrix} \alpha & \delta & \epsilon \\ \delta & \beta & \theta \\ \epsilon & \theta & \gamma \end{bmatrix} . \tag{17.3}$$

Referred to this triad $\boldsymbol{A}$ has the form $(1, i, 0)^T$. Now by (17.1),

$$\begin{pmatrix} \alpha & \delta & \epsilon \\ \delta & \beta & \theta \\ \epsilon & \theta & \gamma \end{pmatrix} \begin{pmatrix} 1 \\ i \\ 0 \end{pmatrix} = \lambda \begin{pmatrix} 1 \\ i \\ 0 \end{pmatrix} . \tag{17.4}$$

Thus

$$\alpha + i\delta = \lambda \,, \quad \delta + i\beta = \lambda i \,, \quad \epsilon + i\theta = 0 \,, \tag{17.5}$$

and hence

$$\widehat{\boldsymbol{Q}} = \begin{pmatrix} \lambda - i\delta & \delta & -i\theta \\ \delta & \lambda + i\delta & \theta \\ -i\theta & \theta & \gamma \end{pmatrix} . \tag{17.6}$$

It is easily checked that $\widehat{\boldsymbol{Q}}$ has the eigenvalues γ and λ (double).

As far as (in)homogeneous waves are considered, this result and its converse mean that if circularly polarised waves propagate for given $\boldsymbol{C}$ or given $\boldsymbol{n}$ (in the case of homogeneous waves), the secular equation must have at least a double root. Conversely if the secular equation has a double root for prescribed $\boldsymbol{C}$ (or $\boldsymbol{n}$) then circularly polarised waves may propagate for that choice of $\boldsymbol{C}$ (or $\boldsymbol{n}$).

18. Isotropic media

We consider trains of inhomogeneous plane waves in the context of classical linear isotropic elasticity theory.

Insertion of (16.1) into (3.2) leads to the propagation condition

$$\{\mu \boldsymbol{C}\cdot\boldsymbol{C}\delta_{ik} + (\lambda+\mu)C_iC_k - \rho N^{-2}\delta_{ik}\}A_k = 0\ , \tag{18.1}$$

and the secular equation

$$\det\{(\mu\boldsymbol{C}\cdot\boldsymbol{C} - \rho N^{-2})\delta_{ik} + (\lambda+\mu)C_iC_k\} = 0\ . \tag{18.2}$$

Equation (18.2) has the double root

$$\mu\boldsymbol{C}\cdot\boldsymbol{C} = \rho N^{-2}\ , \quad \boldsymbol{A}\cdot\boldsymbol{S} = 0\ , \tag{18.3}$$

where $\boldsymbol{A}$ is the corresponding amplitude bivector. This corresponds to the transverse or S - wave. The other root is

$$(\lambda+2\mu)\boldsymbol{C}\cdot\boldsymbol{C} = \rho N^{-2}\ , \quad \boldsymbol{B} = \alpha\boldsymbol{S}\ , \tag{18.4}$$

where $\boldsymbol{B}$ is the corresponding amplitude bivector and α is some scalar. This corresponds to the longitudinal or P - wave.

Because λ and μ are real, it follows from $(18.3)_1$, and $(18.4)_1$, that for both wave solutions, $\boldsymbol{S}^+\cdot\boldsymbol{S}^- = 0$, so that the planes of constant amplitude are orthogonal to the planes of constant phase.

From $(18.3)_1$, we have

$$\mu T^2 e^{2i\phi}(m^2-1) = \rho\ , \tag{18.5}$$

and hence, assuming $m^2 > 1$, without loss in generality, it follows, because T is real, that

$$\phi = 0, \pi, \quad \mu T^2(m^2-1) = \rho, \quad \boldsymbol{S} = \pm(\rho/[\mu(m^2-1)])^{\frac{1}{2}}(m\widehat{\boldsymbol{m}} + i\widehat{\boldsymbol{n}})\ , \tag{18.6}$$

$$|\boldsymbol{S}^-| = (\rho/[\mu(m^2-1)])^{\frac{1}{2}}\ , \quad |\boldsymbol{S}^+| = m(\rho/[\mu(m^2-1)])^{\frac{1}{2}}\ .$$

Clearly $m^2 = 1$ is not possible because $\rho, \mu, T \neq 0$. Because $\boldsymbol{A}$ is any bivector satisfying $\boldsymbol{A}\cdot\boldsymbol{S} = 0$, we have $\boldsymbol{A} = \beta(m^{-1}\widehat{\boldsymbol{m}} + i\widehat{\boldsymbol{n}} + \gamma\widehat{\boldsymbol{p}})$ and the displacement field $\boldsymbol{u}$ is given by

$$\begin{aligned}\boldsymbol{u} = {}& \beta(m^{-1}\widehat{\boldsymbol{m}} + i\widehat{\boldsymbol{n}} + \gamma\widehat{\boldsymbol{p}})\exp i\omega\{\pm(\rho/[\mu(m^2-1)])^{\frac{1}{2}}m\widehat{\boldsymbol{m}}\cdot\boldsymbol{x} - t\}\times \\ & \exp(\mp\omega(\rho/[\mu(m^2-1)])^{\frac{1}{2}}\widehat{\boldsymbol{n}}\cdot\boldsymbol{x})\ ,\end{aligned} \tag{18.7}$$

where β, γ are arbitrary scalars and $\widehat{\boldsymbol{p}}$ is a unit vector orthogonal to the plane of the ellipse of **S**.

The arbitrary scalar γ may be chosen so that this wave is circularly polarised, in which case $\boldsymbol{A}$ is isotropic:

$$(m^{-1}\widehat{\boldsymbol{m}} + i\widehat{\boldsymbol{n}} + \gamma\widehat{\boldsymbol{p}}) \cdot (m^{-1}\widehat{\boldsymbol{m}} + i\widehat{\boldsymbol{n}} + \gamma\widehat{\boldsymbol{p}}) = 0 . \tag{18.8}$$

Then

$$\gamma = \pm((m^2 - 1)/m^2)^{\frac{1}{2}} , \tag{18.9}$$

and

$$\boldsymbol{A} = \beta\{m^{-1}\widehat{\boldsymbol{m}} + i\widehat{\boldsymbol{n}} \pm [(m^2 - 1)/m^2]^{\frac{1}{2}}\widehat{\boldsymbol{p}}\} , \tag{18.10}$$

with $\boldsymbol{S}$ given by (18.6). As m and $\widehat{\boldsymbol{m}}$ vary, the plane of polarisation always contains $\widehat{\boldsymbol{n}}$. The circles of polarisation are obtained by rotating circles about $\widehat{\boldsymbol{n}}$.

Note, from $(18.3)_2$, that the planes of $\boldsymbol{A}$ and $\boldsymbol{S}$ may not be orthogonal, in general. Also, the ellipse of $\boldsymbol{A}$ when projected onto the plane of the ellipse of $\boldsymbol{S}$, is an ellipse whose aspect ratio is the same as the aspect ratio of the ellipse of $\boldsymbol{S}$, is described in the same sense and whose major axis is perpendicular to the major axis of the ellipse of $\boldsymbol{S}$.

For the "longitudinal" waves, we have, from (18.4)

$$(\lambda + 2\mu)T^2 e^{2i\phi}(m^2 - 1) = \rho , \tag{18.11}$$

and

$$\phi = 0 \text{ or } \pi, \quad (\lambda + 2\mu)T^2(m^2 - 1) = \rho , \quad \boldsymbol{S} = \pm(\rho/[(\lambda + 2\mu)(m^2 - 1)])^{\frac{1}{2}}(m\widehat{\boldsymbol{m}} + i\widehat{\boldsymbol{n}}) . \tag{18.12}$$

Because $\boldsymbol{B} = \alpha\boldsymbol{S}$, it follows for the longitudinal wave that the polarisation ellipse is similar and similarly situated to the ellipse of $\boldsymbol{S}$.

Finally, we note, for given $\boldsymbol{C}$, that $\boldsymbol{A} \cdot \boldsymbol{S} = 0$, $\boldsymbol{A} \cdot \boldsymbol{B} = 0$, $\boldsymbol{B} = \alpha\boldsymbol{S}$.

19. Circularly Polarised Inhomogeneous Waves in Cubic Crystals and Incompressible Isotropic Materials.

Here we consider the propagation of circularly polarised inhomogeneous plane waves in (i) cubic crystals (ii) incompressible isotropic elastic bodies. A complete discussion of such waves in transversely isotropic materials is to be found in Boulanger & Hayes (1993).

(i) For **propagation in a cubic crystal** the equations of motion are given by (3.6). Here we choose the prescribed directional ellipse in the xy - plane and thus we let (Hayes 1984)

$$\begin{gathered}\widehat{\boldsymbol{m}} = m_1\boldsymbol{i} + m_2\boldsymbol{j} \ , \quad \widehat{\boldsymbol{n}} = n_1\boldsymbol{i} + n_2\boldsymbol{j} \ , \\ m_1^2 + m_2^2 = n_1^2 + n_2^2 = 1 \ , \quad m_1n_1 + m_2n_2 = 0 \ .\end{gathered} \tag{19.1}$$

Then the components of $\boldsymbol{Q}(m\widehat{\boldsymbol{m}} + i\widehat{\boldsymbol{n}})$ are

$$\begin{aligned} Q_{11} &= m^2(c_{11}m_1^2 + c_{44}m_2^2) + 2im(c_{11}m_1n_1 + c_{44}m_2n_2) - (c_{11}n_1^2 + c_{44}n_2^2) \ , \\ Q_{12} &= [m^2m_1m_2 + im(m_1n_2 + m_2n_1) - n_1n_2](c_{12} + c_{44}) = Q_{21} \ , \\ Q_{22} &= m^2(c_{44}m_1^2 + c_{11}m_2^2) + 2im(c_{11}m_2n_2 + c_{44}m_1n_1) - (c_{11}n_2^2 + c_{44}n_1^2) \ , \\ Q_{33} &= (m^2 - 1)c_{44} \ , \quad Q_{13} = Q_{33} = Q_{31} = Q_{32} = 0 \ . \end{aligned} \tag{19.2}$$

One root, N^{-2}, of the secular equation is given by

$$\rho N^{-2} = (m^2 - 1)c_{44} \ , \tag{19.3}$$

and the other two are the roots N^{-2} of the quadratic

$$(\rho N^{-2})^2 - (m^2 - 1)(c_{11} + c_{44})N^{-2} + \alpha = 0 \ , \tag{19.4}$$

where

$$\begin{aligned} \alpha = {} & m^4[c_{11}c_{44} + \{(c_{11} - c_{44})^2 - (c_{12} + c_{44})^2\}m_1^2m_2^2] + \\ & + 2im^3[(c_{11} - c_{44})^2 - (c_{12} + c_{44})^2](m_2^2 - m_1^2)m_1n_1 + \\ & + m^2[(c_{12} + c_{44})^2 - (c_{11}^2 + c_{44}^2) + 6\{(c_{11} - c_{44})^2 - (c_{12} + c_{44})^2\}m_1^2n_1^2] + \\ & + 2im[(c_{11} - c_{44})^2 - (c_{12} + c_{44})^2](n_1^2 - n_2^2)m_1n_1 + \\ & + [c_{11}c_{44} + \{(c_{11} - c_{44})^2 - (c_{12} + c_{44})^2\}n_1^2n_2^2] \ . \end{aligned} \tag{19.5}$$

Now the cubic secular equation may have a double root if either (19.3) is also a solution of (19.4) or if (19.4) has a double root.

Equation (19.4) has a double root provided

$$(m^2 - 1)^2(c_{11} + c_{44})^2 = 4\alpha \ . \tag{19.6}$$

The root (19.3) is also a solution of (19.4) provided

$$(m^2 - 1)^2c_{11}c_{44} = \alpha \ . \tag{19.7}$$

We consider (19.6) and (19.7).

In general α is complex and (19.6) or (19.7) may not be satisfied because m has to be real. However α becomes real when either $m_1 n_1 = 0$ or $m_2^2 - m_1^2 = n_1^2 - n_2^2 = 0$. We consider in turn these two cases.

Case (1)

Consider first (19.7). It becomes

$$m^2(c_{11} + c_{12})(c_{12} + 2c_{44} - c_{11}) = 0 . \tag{19.8}$$

Thus $m = 0$ and we retrieve the case of homogeneous waves propagating along the axis Oy. (See §8.2).

Next consider (19.6). Let $m_1 = n_2 = 1$, $n_1 = m_2 = 0$. It becomes

$$[(m^2 + 1)(c_{11} - c_{44})]^2 = 4m^2(c_{12} + c_{44})^2 , \tag{19.9}$$

and hence

$$m^2(c_{11} - c_{44}) \pm 2m(c_{12} + c_{44}) + c_{11} - c_{44} = 0 . \tag{19.10}$$

The condition that the roots for m be real is that the discriminant β (say) given by

$$\beta = 4(c_{11} + c_{12})(c_{12} + 2c_{44} - c_{11}) , \tag{19.11}$$

be non-negative. Referring to the table on pages 278-9 of Musgrave's book (1970) we see that $\beta > 0$ for thirteen of the materials listed (including aluminium, iron, nickel) and $\beta < 0$ for the other thirteen cited (including potassium fluoride, sodium bromide, silver bromide). Incidentally, $\beta = 0$ if $c_{11} = c_{12} + 2c_{44}$, in which case the body is isotropic.

Thus in the cases where $\beta > 0$ there are four possible real values for m. Corresponding to each, a circularly polarised wave may propagate, with

$$2\rho T^{-2} e^{-2i\phi} = (m^2 - 1)(c_{11} + c_{44}) , \tag{19.12}$$

where m is given by one of the four roots of (19.10). Typically, the corresponding displacements are of the form

$$\begin{aligned} &u_1 = a \cos \omega(Tmx - t) \exp -\omega Ty, \quad u_2 = a \sin \omega(Tmx - t) \exp -\omega Ty , \\ &u_3 = 0 , \end{aligned} \tag{19.13}$$

where a is a constant, and m satisfies one of the equations (19.10).

Case (2)

Let $m_1 = n_1 = m_2 = -n_2 = 1/\sqrt{2}$. Consider first (19.7). It becomes

$$(m^2 + 1)^2(c_{11} + c_{12})(c_{12} + 2c_{44} - c_{11}) = 0 \tag{19.14}$$

which is not feasible with real m.

Consider now (19.6). It becomes

$$(m^2+1)^2(c_{12}+c_{44})^2 = 4m^2(c_{11}-c_{44})^2 . \qquad (19.15)$$

Thus

$$m^2(c_{12}+c_{44}) \pm 2m(c_{11}-c_{44}) + c_{12}+c_{44} = 0 . \qquad (19.16)$$

The discriminant is $(-\beta)$ where β is given by (19.11). Thus these circularly polarised waves are possible only in the second class (potassium fluoride, etc.).

Thus in the case when $\beta < 0$ there are four possible real values for m. Corresponding to each a circularly polarised wave may propagate with T, ϕ given by (19.12) and m a root of (19.16). Typically, the corresponding displacements are of the form

$$\begin{aligned} &u_1 = a\cos\omega f \exp -\omega g\ , \quad u_2 = a\sin\omega f \exp -\omega g, \ u_3 = 0\ , \\ &f = Tm(x+y)/\sqrt{2} - t\ , \quad g = T(x-y)/\sqrt{2}\ . \end{aligned} \qquad (19.17)$$

(ii) For **propagation in an incompressible isotropic elastic material** the constitutive equation is

$$t_{ij} = -p\delta_{ij} + \mu(u_{i,j}+u_{j,i})\ , \quad u_{k,k} = 0\ , \qquad (19.18)$$

where μ is a constant. Recalling §8.4, the propagation condition is (Hayes, 1984a)

$$iPS_i + \mu\omega A_i \boldsymbol{S}\cdot\boldsymbol{S} = \rho\omega A_i\ . \qquad (19.19)$$

Now $\boldsymbol{A}\cdot\boldsymbol{S} = 0$ and hence **either** (a)

$$\boldsymbol{S}\cdot\boldsymbol{S} = 0\ , \quad \boldsymbol{A} = \alpha\boldsymbol{S}\ , P = -i\alpha\rho\omega\ , \qquad (19.20)$$

where α is a scalar, **or** (b)

$$\boldsymbol{S}\cdot\boldsymbol{A} = 0\ , \quad P = 0\ , \quad \mu\boldsymbol{S}\cdot\boldsymbol{S} = \rho\ . \qquad (19.21)$$

In case (a) we have the solution

$$\boldsymbol{u} = \{\alpha\boldsymbol{S}\exp i\omega(\boldsymbol{S}\cdot\boldsymbol{x}-t)\}^+\ , \ p = -\{i\rho\omega\alpha\exp i\omega(\boldsymbol{S}\cdot\boldsymbol{x}-t)\}^+\ , \ \boldsymbol{S}\cdot\boldsymbol{S} = 0\ , \qquad (19.22)$$

and α is arbitrary. Choosing coordinate axes $0x, 0y$ along $\boldsymbol{S}^+$ and $\boldsymbol{S}^-$, respectively, and taking α real for simplicity, this solution may be written

$$\begin{aligned} &u = \alpha S\cos\omega(Sx-t)\ , \\ &v = -\alpha S\sin\omega(Sx-t)\ , \quad w = 0\ , \quad p = \rho\omega\alpha\sin\omega(Sx-t)\ , \end{aligned} \qquad (19.23)$$

where $S = |S^+| = |S^-|$, ω and α are arbitrary real numbers. This is a circularly polarised wave. Planes of constant phase are orthogonal to planes of constant amplitude. Also $|S^+| = |S^-|$.

In case (b)

$$\boldsymbol{u} = \boldsymbol{A} \exp i\omega(\boldsymbol{S}\cdot\boldsymbol{x} - t)\ , \quad \mu\boldsymbol{S}\cdot\boldsymbol{S} = \rho\ , \quad p \doteq 0\ , \tag{19.24}$$

where $\boldsymbol{A}$ is any bivector satisfying $\boldsymbol{A}\cdot\boldsymbol{S} = 0$. We have

$$\mu T^2(m^2 - 1) = \rho e^{-2i\phi}\ , \tag{19.25}$$

and so, for $m^2 > 1$,

$$\boldsymbol{S} = \pm(\rho/[\mu(m^2-1)])^{\frac{1}{2}}(m\widehat{\boldsymbol{m}} + i\widehat{\boldsymbol{n}})\ , \quad \boldsymbol{A} = \beta(m^{-1}\widehat{\boldsymbol{m}} + i\widehat{\boldsymbol{n}} + \gamma\widehat{\boldsymbol{p}})\ , \tag{19.26}$$

where $\widehat{\boldsymbol{p}}$ is a unit vector orthogonal to $\widehat{\boldsymbol{m}}$ and $\widehat{\boldsymbol{n}}$, and γ is arbitrary. By choosing $\gamma^2 = 1 - (1/m^2)$, the wave is circularly polarised.

20. Energy flux for Inhomogeneous Plane Waves

We consider the propagation of a single infinite train of elliptically polarised time-harmonic small amplitude inhomogeneous plane waves in a homogeneous conservative system. We try to be as general as possible. (Hayes 1980).

Thus, the propagating field quantity is assumed to have the form

$$\boldsymbol{A} \exp i\omega(\boldsymbol{S}\cdot\boldsymbol{x} - t)\ . \tag{20.1}$$

The partial differential equations with constant coefficients governing the system will lead to a propagation condition of the form

$$\boldsymbol{L}\boldsymbol{A} = 0\ , \tag{20.2}$$

where $\boldsymbol{L}$ is a second order tensor, which, in general, may depend on $\boldsymbol{S}$ and on ω.

Now to the three assumptions of §7, we add a fourth, namely that $\boldsymbol{L}$ depends only upon $\boldsymbol{S}$ and thus not on the angular frequency ω:

$$\boldsymbol{L} = \boldsymbol{L}(\boldsymbol{S}). \tag{20.3}$$

It follows that the secular equation corresponding to the propagation condition (20.2) reads

$$Q(\boldsymbol{S}) = \det \boldsymbol{L}(\boldsymbol{S}) = 0\ , \tag{20.4}$$

where Q is independent of ω. This is certainly the case for linear homogeneous anisotropic elasticity theory. Other cases will suggest themselves. We note that for homogeneous waves the bivector $\boldsymbol{S}$ has to be replaced by the vector $\boldsymbol{S} = (1/\omega)\boldsymbol{k}$, and that (20.4) thus reads $Q(\boldsymbol{k}/\omega) = 0$. It follows that the dispersion relations obtained by solving this equation for ω at given $\boldsymbol{k}$ must then be homogeneous of degree one in $\boldsymbol{k}$.

When an $\boldsymbol{S}$ is obtained which satisfies the secular equation $Q(\boldsymbol{S}) = 0$, the corresponding $\boldsymbol{A}$ is determined from the propagation condition (20.2). Note the amplitude eigenbivector $\boldsymbol{A}$ so obtained is independent of ω, because $\boldsymbol{L}$ is independent of ω. The propagation condition (20.2) is homogeneous and hence any scalar multiple of $\boldsymbol{A}$ is also a suitable amplitude bivector. To accommodate boundary or initial conditions which may involve the frequency ω, the field is assumed to have the form

$$f(\omega)\boldsymbol{A}\exp i\omega(\boldsymbol{S}\cdot\boldsymbol{x} - t)\ . \tag{20.5}$$

Here $f(\omega)$ is an arbitrary function of $\omega, \boldsymbol{S}$ is a solution of the secular equation $Q(\boldsymbol{S}) = 0$, and $\boldsymbol{A}$ is the corresponding eigenbivector of (20.2).

If one field quantity has the form (20.5) then every other field quantity entering the expressions for the energy flux and energy density has the similar form

$$f(\omega)\boldsymbol{B}\exp i\omega(\boldsymbol{S}\cdot\boldsymbol{x} - t)\ , \tag{20.6}$$

where the same function $f(\omega)$ occurs in both expressions and $\boldsymbol{B}$ may be a (complex) scalar, vector or tensor.

The energy flux vector has the **form** given by

$$\boldsymbol{r} = \{f(\omega)\boldsymbol{A}\exp i\omega(\boldsymbol{S}\cdot\boldsymbol{x} - t) + c\cdot c\} \times \{f(\omega)\boldsymbol{B}\exp i\omega(\boldsymbol{S}\cdot\boldsymbol{x} - t) + c\cdot c\}\ , \tag{20.7}$$

so that

$$\boldsymbol{r} = [\{f(\omega)\}^2\boldsymbol{T}\exp[2i\omega(\boldsymbol{S}\cdot\boldsymbol{x} - t)] + c\cdot c] + f(\omega)\overline{f(\omega)}\boldsymbol{h}\exp[i\omega(\boldsymbol{S} - \overline{\boldsymbol{S}})\cdot\boldsymbol{x}]\ , \tag{20.8}$$

where $\boldsymbol{T}$ is a bivector and $\boldsymbol{h}$ is a real vector, both independent of ω. [See the example of linear homogeneous anisotropic elasticity theory (§6).] Here c.c. stands for complex conjugate.

Similarly, the energy density, e, has the form

$$e = [\{f(\omega)\}^2 B\exp[2i\omega(\boldsymbol{S}\cdot\boldsymbol{x} - t)] + c\cdot c] + f(\omega)\overline{f(\omega)}\beta\exp[i\omega(\boldsymbol{S} - \overline{\boldsymbol{S}})\cdot\boldsymbol{x}]\ , \tag{20.9}$$

where B is a complex constant and β is a real constant, both independent of ω.

Taking the mean over a period, we note that

$$\begin{aligned} \tilde{\boldsymbol{r}} &= f\overline{f}\boldsymbol{h}\exp[i\omega(\boldsymbol{S} - \overline{\boldsymbol{S}})\cdot\boldsymbol{x}] = f\overline{f}\boldsymbol{h}\exp(-2\omega\boldsymbol{S}^-\cdot\boldsymbol{x})\ , \\ \tilde{e} &= f\overline{f}\beta\exp[i\omega(\boldsymbol{S} - \overline{\boldsymbol{S}})\cdot\boldsymbol{x}] = f\overline{f}\beta\exp(-2\omega\boldsymbol{S}^-\cdot\boldsymbol{x})\ . \end{aligned} \tag{20.10}$$

In contrast with the case of homogeneous waves, the mean energy flux and mean energy density depend upon position. This is due to the spatial attentuation of the inhomogeneous waves.

For further reference we introduce here the "weighted mean" energy flux vector $\hat{\boldsymbol{r}}$ and the "weighted mean" energy density $\hat{e}$. These are defined by

$$\begin{aligned}\hat{\boldsymbol{r}} &= \tilde{\boldsymbol{r}} \exp(2\omega \boldsymbol{S}^- \cdot \boldsymbol{x}) = f\overline{f}\boldsymbol{h} \,, \\ \hat{e} &= \tilde{e} \exp(2\omega \boldsymbol{S}^- \cdot \boldsymbol{x}) = f\overline{f}\beta \,,\end{aligned} \tag{20.11}$$

and are constant for a given inhomogeneous wave solution.

Now inserting (20.8) and (20.9) into the conservation of energy equation (6.6) which has to be satisfied for all $\boldsymbol{x}$ and t, we conclude that

$$\boldsymbol{h} \cdot (\boldsymbol{S} - \overline{\boldsymbol{S}}) = 0 \,, \tag{20.12}$$

or, equivalently,

$$\tilde{\boldsymbol{r}} \cdot \boldsymbol{S}^- = 0 \,. \tag{20.13}$$

Thus the mean energy flux vector is parallel to the planes of constant amplitude, $\boldsymbol{S}^- \cdot \boldsymbol{x} =$ constant.

Because the solutions are valid for arbitrary ω, they are equally valid when the real ω is replaced by a complex $\omega^\star$, whose real part is ω, the real frequency of the waves under consideration. Then $\boldsymbol{r}$, given by (20.8), is replaced by $\boldsymbol{r}(\omega^\star)$, say, given by

$$\begin{aligned}\boldsymbol{r}(\omega^\star) &= \{[f(\omega^\star)]^2 \boldsymbol{T} \exp[2i\omega^\star(\boldsymbol{S}\cdot\boldsymbol{x} - t)] + c\cdot c\} \\ &\quad + f(\omega^\star)\overline{f(\omega^\star)}\boldsymbol{h} \exp[i\omega^\star(\boldsymbol{S}\cdot\boldsymbol{x} - t) - \overline{i\omega^\star}(\overline{\boldsymbol{S}}\cdot\boldsymbol{x} - t)] \,.\end{aligned} \tag{20.14}$$

Here $\boldsymbol{T}$ and $\boldsymbol{h}$ are unchanged because they are independent of ω. Similarly, the energy density, e, is replaced by $e(\omega^\star)$, say, given by

$$\begin{aligned}e(\omega^\star) &= \{[f(\omega^\star)]^2 B \exp[2i\omega^\star(\boldsymbol{S}\cdot\boldsymbol{x} - t)] + c\cdot c\} \\ &\quad + f(\omega^\star)\overline{f(\omega^\star)}\beta \exp[i\omega^\star(\boldsymbol{S}\cdot\boldsymbol{x} - t) - \overline{i\omega^\star}(\overline{\boldsymbol{S}}\cdot\boldsymbol{x} - t)] \,.\end{aligned} \tag{20.15}$$

The energy conservation equation (6.6) now gives

$$\boldsymbol{h} \cdot (\omega^\star \boldsymbol{S} - \overline{\omega^\star \boldsymbol{S}}) = \beta(\omega^\star - \overline{\omega^\star}) \,. \tag{20.16}$$

Using (20.12) it follows that

$$\boldsymbol{h} \cdot \boldsymbol{S} = \beta \,, \tag{20.17}$$

or equivalently

$$\tilde{\boldsymbol{r}} \cdot \boldsymbol{S} = \tilde{e} \,. \tag{20.18}$$

Hence, taking real and imaginary parts,

$$\tilde{\boldsymbol{r}} \cdot \boldsymbol{S}^+ = \tilde{e}, \;\; \tilde{\boldsymbol{r}} \cdot \boldsymbol{S}^- = 0 \, . \tag{20.19}$$

Thus the component of the mean energy flux vector along the normal to the planes of constant phase is equal to the phase speed $(1/|\boldsymbol{S}^+|)$ times the mean energy density. Using (7.5) we have, for the energy flux velocity $\boldsymbol{g}$,

$$\boldsymbol{g} \cdot \boldsymbol{S}^+ = 1, \;\; \boldsymbol{g} \cdot \boldsymbol{S}^- = 0 \, . \tag{20.20}$$

Of course if $\boldsymbol{S}^- = 0$, so that the waves are not attentuated, then the result $\boldsymbol{g} \cdot \boldsymbol{S}^+ = 1$ is still valid.

21. Example: Energy Flux for Inhomogeneous Waves in Isotropic Media

For the wave train (16.1) propagating in an isotropic medium, we have

$$v_i = \{i\omega A_i \exp i\omega(\boldsymbol{S} \cdot \boldsymbol{x} - t)\}^+ \, , \tag{21.1}$$

$$e_{ij} = (1/2)\{i\omega(A_i S_j + A_j S_i) \exp i\omega(\boldsymbol{S} \cdot \boldsymbol{x} - t)\}^+ \, , \tag{21.2}$$

$$t_{ij} = \{i\omega[\lambda(\boldsymbol{S} \cdot \boldsymbol{A})\delta_{ij} + \mu(A_i S_j + A_j S_i)] \exp i\omega(\boldsymbol{S} \cdot \boldsymbol{x} - t)\}^+ \, . \tag{21.3}$$

Thus, recalling (6.2), (6.4) and (6.7), we obtain for the weighted mean energy flux and weighted mean energy density

$$4\hat{\boldsymbol{r}}/\omega^2 = \lambda\{(\boldsymbol{S} \cdot \boldsymbol{A})\overline{\boldsymbol{A}} + (\overline{\boldsymbol{S}} \cdot \overline{\boldsymbol{A}})\boldsymbol{A}\} + \mu\{(\boldsymbol{A} \cdot \overline{\boldsymbol{A}})(\boldsymbol{S} + \overline{\boldsymbol{S}}) + (\boldsymbol{S} \cdot \overline{\boldsymbol{A}})\boldsymbol{A} + (\overline{\boldsymbol{S}} \cdot \boldsymbol{A})\overline{\boldsymbol{A}}\} \, , \tag{21.4}$$

$$4\hat{e}/\omega^2 = \rho \boldsymbol{A} \cdot \overline{\boldsymbol{A}} + \lambda(\boldsymbol{S} \cdot \boldsymbol{A})(\overline{\boldsymbol{S}} \cdot \overline{\boldsymbol{A}}) + \mu\{\boldsymbol{A} \cdot \overline{\boldsymbol{A}})(\boldsymbol{S} \cdot \overline{\boldsymbol{S}}) + (\boldsymbol{S} \cdot \overline{\boldsymbol{A}})(\overline{\boldsymbol{S}} \cdot \boldsymbol{A})\} \, . \tag{21.5}$$

Now, taking the dot product of (21.4) with the slowness bivector $\boldsymbol{S}$, and using (21.5) yields

$$\hat{\boldsymbol{r}} \cdot \boldsymbol{S} = \hat{e} + \frac{\omega^2}{4}\{(\boldsymbol{A} \cdot \overline{\boldsymbol{A}})(\mu \boldsymbol{S} \cdot \boldsymbol{S} - \rho) + (\lambda + \mu)(\boldsymbol{S} \cdot \boldsymbol{A})(\boldsymbol{S} \cdot \overline{\boldsymbol{A}})\} \, . \tag{21.6}$$

For the transverse wave (18.3), $\boldsymbol{S} \cdot \boldsymbol{A} = 0$ and $\mu \boldsymbol{S} \cdot \dot{\boldsymbol{S}} = \rho$, so that this reduces to $\hat{\boldsymbol{r}} \cdot \boldsymbol{S} = \hat{e}$. For the longitudinal wave (18.4), $\boldsymbol{A} = \alpha \boldsymbol{S}$ and $(\lambda + 2\mu)\boldsymbol{S} \cdot \boldsymbol{S} = \rho$, so that (21.6) again reduces to $\widehat{\boldsymbol{r}} \cdot \boldsymbol{S} = \widehat{e}$. Thus, in both cases, (20.18), or equivalently (20.19), is confirmed.

For the transverse wave, introducing $\boldsymbol{S}$ given by (18.6), and $\boldsymbol{A}$ given by (18.7), into (21.4), (21.5) yields

$$\widehat{e} = |\beta|^2(\widehat{e}_1 + |\gamma|^2\widehat{e}_2) \, , \tag{21.7}$$

$$\widehat{\boldsymbol{r}} = |\beta|^2(\widehat{\boldsymbol{r}}_1 + |\gamma|^2\widehat{\boldsymbol{r}}_2 + \gamma^+\widehat{\boldsymbol{r}}_{12}) \, , \tag{21.8}$$

with

$$\widehat{e}_1 = (\rho\omega^2/2)\frac{m^2+3}{m^2-1} \quad , \quad \widehat{e}_2 = (\rho\omega^2/2)\frac{m^2}{m^2-1} \, , \tag{21.9}$$

and

$$\widehat{\boldsymbol{r}}_1 = \pm(\rho\omega^2/2)\sqrt{\frac{\mu}{\rho(m^2-1)}}\frac{m^2+3}{m}\widehat{\boldsymbol{m}} \quad , \quad \widehat{\boldsymbol{r}}_2 = \pm(\rho\omega^2/2)\sqrt{\frac{\mu}{\rho(m^2-1)}}m\widehat{\boldsymbol{m}} \, , \tag{21.10}$$

$$\widehat{\boldsymbol{r}}_{12} = \pm\rho\omega^2\sqrt{\frac{\mu}{\rho(m^2-1)}}\widehat{\boldsymbol{p}} \, . \tag{21.11}$$

Thus the energy flux vector always lies in the plane of $\widehat{\boldsymbol{m}}$ and $\widehat{\boldsymbol{p}}$. Its direction varies when γ is varied. When $\beta = 1, \gamma = 0$ (ellipical polarisation in the plane of the slowness bivector), $\widehat{\boldsymbol{r}} = \widehat{\boldsymbol{r}}_1$ and $\widehat{e} = \widehat{e}_1$; when $\beta \to 0$, $\gamma \to \infty$ with $\beta\gamma = 1$ (linear polarisation in the direction orthogonal to the plane of the slowness bivector), $\widehat{\boldsymbol{r}} = \widehat{\boldsymbol{r}}_2$ and $\widehat{e} = \widehat{e}_2$. For these two special cases the energy flux velocity is equal to the phase speed :

$$\boldsymbol{g}_1 = \widehat{\boldsymbol{r}}_1/\widehat{e}_1 = \boldsymbol{g}_2 = \widehat{\boldsymbol{r}}_2/\widehat{e}_2 = \pm\sqrt{(\mu/\rho)(m^2-1)}m^{-1}\widehat{\boldsymbol{m}} \, . \tag{21.12}$$

However, in general, the energy flux velocity has also a component orthogonal to the plane of the slowness bivector.

For the longitudinal wave, introducing $\boldsymbol{S}$ given by (18.12), and $\boldsymbol{A} = \alpha\boldsymbol{S}$, into (21.4) and (21.5) yields

$$\widehat{e} = (\rho\omega^2/2)|\alpha|^2m^2\frac{\lambda(m^2-1)+2\mu(m^2+1)}{(\lambda+2\mu)(m^2-1)} \, , \tag{21.13}$$

$$\widehat{\boldsymbol{r}} = \pm(\rho\omega^2/2)|\alpha|^2m\frac{\lambda(m^2-1)+2\mu(m^2+1)}{\sqrt{\rho(\lambda+2\mu)(m^2-1)}}\widehat{\boldsymbol{m}} \, . \tag{21.14}$$

Thus, for the longitudinal wave, the energy flux is along $\widehat{\boldsymbol{m}}$. The energy flux velocity is equal to the phase speed :

$$\boldsymbol{g} = \widehat{\boldsymbol{r}}/\widehat{e} = \pm\sqrt{\{(\lambda+2\mu)/\rho\}(m^2-1)}m^{-1}\widehat{\boldsymbol{m}} \, . \tag{21.15}$$

22. Two or More Wave Trains

Because the field components combine in pairs in the formation of the energy flux vector and the energy density, it is sufficient to consider just two wave trains to obtain results which are valid for two or more wave trains.

Suppose the trains have the same angular frequency ω, and have slownesses $\boldsymbol{S}_1$ and $\boldsymbol{S}_2$. Then the energy flux vector for the superposition of the two trains has the form

$$\begin{aligned} \boldsymbol{r} &= \{f_1(\omega)\boldsymbol{A}_1 \exp i\omega(\boldsymbol{S}_1\cdot\boldsymbol{x}-t) + f_2(\omega)\boldsymbol{A}_2 \exp i\omega(\boldsymbol{S}_2\cdot\boldsymbol{x}-t) + c\cdot c\}\times \\ &\quad \{f_1(\omega)\boldsymbol{B}_1 \exp i\omega(\boldsymbol{S}_1\cdot\boldsymbol{x}-t) + f_2(\omega)\boldsymbol{B}_2 \exp i\omega(\boldsymbol{S}\cdot\boldsymbol{x}-t) + c\cdot c\} \\ &= \sum_{\alpha=1}^{2}\{f_\alpha^2(\omega)\boldsymbol{T}_\alpha \exp 2i\omega(\boldsymbol{S}_\alpha\cdot\boldsymbol{x}-t) + c\cdot c\} \\ &\quad + \{f_1 f_2 \boldsymbol{T}_{12} \exp i\omega[(\boldsymbol{S}_1+\boldsymbol{S}_2)\cdot\boldsymbol{x}-2t] + c\cdot c\} + \tilde{\boldsymbol{r}}\,, \end{aligned} \tag{22.1}$$

where $\boldsymbol{A}_\alpha, \boldsymbol{B}_\alpha$ may be complex scalars, vectors or tensors, independent of ω. Also $\boldsymbol{T}_\alpha, \boldsymbol{T}_{12}$ are bivectors independent of ω; $f_1(\omega)$ and $f_2(\omega)$ are arbitrary functions of ω, and $\tilde{\boldsymbol{r}}$ is the mean of $\boldsymbol{r}$, given by

$$\tilde{\boldsymbol{r}} = \sum_{\alpha=1}^{2} f_\alpha \overline{f}_\alpha \boldsymbol{w}_\alpha \exp i\omega(\boldsymbol{S}_\alpha - \overline{\boldsymbol{S}}_\alpha)\cdot\boldsymbol{x} + [f_1\overline{f}_2 \boldsymbol{D}_{12} \exp i\omega(\boldsymbol{S}_1 - \overline{\boldsymbol{S}}_2)\cdot\boldsymbol{x} + c\cdot c]\,. \tag{22.2}$$

Here $\boldsymbol{w}_\alpha$ are real vectors and $\boldsymbol{D}_{12}$ is a bivector, all independent of ω.

Similarly, for the energy density of the combined wave trains, we have

$$e = \sum_{\alpha=1}^{2}\{f_\alpha^2 B_\alpha \exp 2i\omega(\boldsymbol{S}_\alpha\cdot\boldsymbol{x}-t)+c\cdot c\}+[f_1\overline{f}_2 B_{12} \exp\{i\omega(\boldsymbol{S}_1+\overline{\boldsymbol{S}}_2)\cdot\boldsymbol{x}-2t\}+c\cdot c]+\tilde{e}\,, \tag{22.3}$$

where B_α, B_{12} are complex scalars independent of ω, and $\tilde{e}$ is the mean of e, given by

$$\tilde{e} = \sum_{\alpha=1}^{2} f_\alpha \overline{f}_\alpha \gamma_\alpha \exp i\omega(\boldsymbol{S}_\alpha - \overline{\boldsymbol{S}}_\alpha)\cdot\boldsymbol{x} + [f_1\overline{f}_2 G_{12} \exp i\omega(\boldsymbol{S}_1 - \overline{\boldsymbol{S}}_2)\cdot\boldsymbol{x} + c\cdot c]\,. \tag{22.4}$$

Here γ_α are real scalars, and G_{12} is a complex scalar, all independent of ω.

Insertion of (22.1) and (22.3) into the energy conservation equation (6.6) gives

$$\boldsymbol{w}_1\cdot(\boldsymbol{S}_1 - \overline{\boldsymbol{S}}_1) = \boldsymbol{w}_2\cdot(\boldsymbol{S}_2 - \overline{\boldsymbol{S}}_2) = \boldsymbol{D}_{12}\cdot(\boldsymbol{S}_1 - \overline{\boldsymbol{S}}_2) = 0\,. \tag{22.5}$$

As before, if ω is replaced by the complex $\omega^\star$ whose real part is ω, the real frequency of the wave trains under consideration, then $\tilde{\boldsymbol{r}}$, given by (22.2), is replaced by $\tilde{\boldsymbol{r}}(\omega^\star)$, say, given by

$$\tilde{\boldsymbol{r}}(\omega^\star) = \sum_{\alpha=1}^{2} f_\alpha(\omega^\star)\overline{f_\alpha(\omega^\star)}\boldsymbol{w}_\alpha \exp[i\omega^\star(\boldsymbol{S}_\alpha \cdot \boldsymbol{x} - t) - i\overline{\omega^\star}(\overline{\boldsymbol{S}}_\alpha \cdot \boldsymbol{x} - t)]$$
$$+ \{f_1(\omega^\star)\overline{f_2(\omega^\star)}\boldsymbol{D}_{12} \exp[i\omega^\star(\boldsymbol{S}_1 \cdot \boldsymbol{x} - t) - i\overline{\omega^\star}(\overline{\boldsymbol{S}}_2 \cdot \boldsymbol{x} - t)] + c \cdot c\} \ . \tag{22.6}$$

Also, $\tilde{e}$ is replaced by $\tilde{e}(\omega^\star)$, say, given by

$$\tilde{e}(\omega^\star) = \sum_{\alpha=1}^{2} f_\alpha(\omega^\star)\overline{f_\alpha(\omega^\star)}\gamma_\alpha \exp[i\omega^\star(\boldsymbol{S}_\alpha \cdot \boldsymbol{x} - t) - i\overline{\omega^\star}(\overline{\boldsymbol{S}}_\alpha \cdot \boldsymbol{x} - t)]$$
$$+ \{f_1(\omega^\star)\overline{f_2(\omega^\star)}G_{12} \exp[i\omega^\star(\boldsymbol{S}_1 \cdot \boldsymbol{x} - t) - i\overline{\omega^\star}(\overline{\boldsymbol{S}}_2 \cdot \boldsymbol{x} - t)] + c \cdot c\} \ . \tag{22.7}$$

Now the energy conservation equation (6.6) gives

$$\begin{aligned} \boldsymbol{w}_1 \cdot (\omega^\star \boldsymbol{S}_1 - \overline{\omega^\star \boldsymbol{S}_1}) &= \gamma_1(\omega^\star - \overline{\omega}^\star) \ , \\ \boldsymbol{w}_2 \cdot (\omega^\star \boldsymbol{S}_2 - \overline{\omega^\star \boldsymbol{S}_2}) &= \gamma_2(\omega^\star - \overline{\omega}^\star) \ , \\ \boldsymbol{D}_{12} \cdot (\omega^\star \boldsymbol{S}_1 - \overline{\omega^\star \boldsymbol{S}_2}) &= G_{12}(\omega^\star - \overline{\omega}^\star) \ . \end{aligned} \tag{22.8}$$

Thus, by using (4.5), it follows that

$$\boldsymbol{w}_1 \cdot \boldsymbol{S}_1 = \gamma_1, \quad \boldsymbol{w}_2 \cdot \boldsymbol{S}_2 = \gamma_2 \ , \quad \boldsymbol{D}_{12} \cdot \boldsymbol{S}_1 = \boldsymbol{D}_{12} \cdot \overline{\boldsymbol{S}}_2 = G_{12} \ . \tag{22.9}$$

These results are valid for two wave trains which have the same frequency but different slownesses $\boldsymbol{S}_1$ and $\boldsymbol{S}_2$. Similar results are valid for more than two trains.

These results are equally valid for two or more wave trains in which the frequency ω is no longer real but complex.

We present some examples.

Example 1. Surface Waves.

Suppose that two plane waves with slownesses $\boldsymbol{S}_1$ and $\boldsymbol{S}_2$ are propagating over the plane surface $\boldsymbol{m} \cdot \boldsymbol{x} = 0$.

In order that the amplitude decay with distance from the surface, $\boldsymbol{S}_1^-$ and $\boldsymbol{S}_2^-$ are both parallel to $\boldsymbol{m}$. Some boundary condition is imposed on the surface and hence the waves must be in phase on $\boldsymbol{m} \cdot \boldsymbol{x} = 0$. Thus $\boldsymbol{S}_1^+$ and $\boldsymbol{S}_2^+$ must have the form

$$\boldsymbol{S}_1^+ = \boldsymbol{a}/c + p_1\boldsymbol{m}, \quad \boldsymbol{S}_2^+ = \boldsymbol{a}/c + p_2\boldsymbol{m}, \quad \boldsymbol{a} \cdot \boldsymbol{m} = 0 \ , \tag{22.10}$$

where $(1/c)$ is the common in-surface component of the slownesses and p_1, p_2 are real constants. Thus $(\boldsymbol{S}_1 - \overline{\boldsymbol{S}}_1)$, $(\boldsymbol{S}_2 - \overline{\boldsymbol{S}}_2)$, and $(\boldsymbol{S}_1 - \overline{\boldsymbol{S}}_2)$ are all parallel to $\boldsymbol{m}$. Hence, using (22.5),

$$\boldsymbol{w}_1 \cdot \boldsymbol{m} = \boldsymbol{w}_2 \cdot \boldsymbol{m} = \boldsymbol{D}_{12} \cdot \boldsymbol{m} = 0 \ . \tag{22.11}$$

Then, using (22.9),

$$\boldsymbol{w}_1 \cdot \boldsymbol{a} = \gamma_1 c, \quad \boldsymbol{w}_2 \cdot \boldsymbol{a} = \gamma_2 c, \quad \boldsymbol{D}_{12} \cdot \boldsymbol{a} = G_{12} c \,. \tag{22.12}$$

Hence, using (22.2) and (22.4),

$$\tilde{\boldsymbol{r}} \cdot \boldsymbol{a} = c\tilde{e} \,, \quad \tilde{\boldsymbol{r}} \cdot \boldsymbol{m} = 0 \,. \tag{22.13}$$

Here $\tilde{\boldsymbol{r}}$ and $\tilde{e}$ are the mean energy flux and energy density for the combined wave motion. Using (6.5), we have, for the energy flux velocity of the combined motion,

$$\boldsymbol{g} \cdot \boldsymbol{a} = c, \quad \boldsymbol{g} \cdot \boldsymbol{m} = 0 \,. \tag{22.14}$$

Example 2. Homogeneous Waves with a Common Propagation Direction. Suppose q wave trains with real slownesses propagate in the same direction. Thus, $\boldsymbol{S}_\alpha = s_\alpha \boldsymbol{n}, \quad \alpha = 1, \ldots q$. By analogy with (22.2) and (22.4), the mean energy flux vector and mean energy density corresponding to the superposition of these waves are given by

$$\begin{aligned} \tilde{\boldsymbol{r}} &= \sum_{\alpha=1}^{q} f_\alpha \bar{f}_\alpha \boldsymbol{w}_\alpha + \sum_{\alpha=1}^{q} \sum_{\beta=1}^{q} \{ f_\alpha \bar{f}_\beta \boldsymbol{D}_{\alpha\beta} \exp i\omega(s_\alpha - s_\beta) \boldsymbol{n} \cdot \boldsymbol{x} + c \cdot c \} \,, \\ \tilde{e} &= \sum_{\alpha=1}^{q} f_\alpha \bar{f}_\alpha \gamma_\alpha + \sum_{\alpha=1}^{q} \sum_{\beta=1}^{q} \{ f_\alpha \bar{f}_\beta G_{\alpha\beta} \exp i\omega(s_\alpha - s_\beta) \boldsymbol{n} \cdot \boldsymbol{x} + c \cdot c \} \,. \end{aligned} \tag{22.15}$$

Using (22.5) and (22.9) we have

$$\boldsymbol{D}_{\alpha\beta} \cdot \boldsymbol{n} = 0 \,, \quad G_{\alpha\beta} = 0 \,, \quad s_\alpha \boldsymbol{w}_\alpha \cdot \boldsymbol{n} = \gamma_\alpha \,, \tag{22.16}$$

and hence,

$$\tilde{\boldsymbol{r}} \cdot \boldsymbol{n} = \sum_{\alpha=1}^{q} \tilde{e}_\alpha / s_\alpha \,. \tag{22.17}$$

Here $\tilde{\boldsymbol{r}}$ is the mean energy flux for the resultant motion and $\tilde{e}_\alpha$ is the mean energy density for the individual motion with slowness $s_\alpha \boldsymbol{n}$. We note that $G_{\alpha\beta} = 0$, which means that the mean energy density of the resultant motion is equal to the sum of the mean energy densities of the individual motions.

References

Boulanger, Ph. and M. Hayes, 1991 Quart. J. Mech. Appl. Math., **44**, 235-240; 1993 *Bivectors and Waves in Mechanics and Optics*, Chapman- Hall, London.

Browder, F.E. 1954 Ann. Math. Stud., **33**, 15-51.

Fedorov, F.I. 1968 *Theory of Elastic Waves in Crystals*, Plenum Press, New York.

Gibbs, J.W. 1881, 1884 Elements of Vector Analysis (privately printed) ≡ pp 17-90, Vol 2, *Scientific Papers*, Dover, New York 1961.

Hamilton, W.R. 1853 *Lectures on Quaternions*, Hodges & Smith, Dublin.

Hayes, M. 1963 Proc. Roy. Soc. London. A **274**, 500-506; 1972 Arch. Rat'l Mech. Anal. **46**, 105-113; 1979 Proc. Roy. Irish Acad. **79**, 15-28; 1980 Proc. Roy. Soc. London A **370**, 417-429; 1984 Arch. Rat'l Mech. Anal. **85**, 41-79; 1984a in Wave Phenomena: *Modern Theory and Applications* (ed. C. Rogers and T.B. Moodie), Elsevier 175-191.

Kelvin, Lord 1904 *Baltimore Lectures*, Cambridge Univesity Press.

Klerk, J. de and M.J.P. Musgrave 1955 Proc. Phys. Soc., B, **68**, 81-88.

Love, A.E.H. 1927 *A treatise on the Mathematical Theory of Elasticity*, 4th ed., University Press, Cambridge.

MacCullagh, J. 1847 Proc. Roy. Irish. Acad., **3**, 49-51.

Morrey, C.B. 1954 Ann. Math. Stud., **33**, 101-159.

Musgrave, M.J.P. 1954 Proc. Roy. Soc. London. A, **226**, 339-366; 1970 *Crystal Acoustics*, Holden-Day, San Francisco.

Stone, J.M. 1963 *Radiation and Optics*, McGraw-Hill, New York.

Synge, J.L. 1957 J. Maths and Phys. **35**, 323-334; 1964 The Petrov classification of gravitational fields. Comm. Dublin Inst. for Advanced Studies, A, **15**.

STABILITY OF FLOW IN A DIVERGING CHANNEL

Ph.G. Drazin
University of Bristol, Bristol, UK

Abstract

The linear, weakly nonlinear, and nonlinear stability of flows of a viscous incompressible fluid in a diverging channel will be treated theoretically. The results will be related to observations of flows, to computational fluid dynamics, to bifurcations of the solutions describing the flows as the Reynolds number increases, and to transition towards chaos. This will also serve as a case study of the concepts and methods of the theory of stability of all flows.

1. Flow in a Diverging Channel

The flow of an incompressible viscous fluid driven steadily along a two-dimensional channel poses a classic problem of fluid mechanics. Its study goes back at least as far as the fifteenth century, when Leonardo da Vinci was commissioned by Duke Ludovico Sforza of Milano and by the Signoria of Firenze to engineer their canals. In modern times the theory of flow in a diverging channel has many applications in aerospace, chemical, civil, environmental, and mechanical engineering as well as in understanding rivers and canals. It is discussed in many textbooks [e.g. 1, §5.6]. It has been studied by many persons using mathematical, numerical and experimental methods. Their results appear in many papers. However, some fundamental aspects of the problem are still poorly understood. The problem will be reviewed in this course of five lectures. The concepts and methods which we will use are important in the study of the stability of all flows. In this first lecture we shall set the scene by looking at flows in channels of various shapes and sizes at various speeds, and then develop some general concepts as a basis to understand what happens and why.

Many excellent photographs of flows in channels (and pipes) have been published by Nakayama [2, Figs. 105–115]. It is valuable to ponder over these photographs. Several points appear soon to the observant reader. Separation of the flow from a wall, and the consequent reversal of flow may be noted. The difference between flows in converging and diverging channels is very marked. The difference between flows along a slowly divergent channel and a sudden expansion (i.e. a channel with horizontal walls having a discontinuous increase of height at some station, seen in, e.g., [2, Fig. 113]) is marked. Nakayama does not make it entirely clear which photographs are of axisymmetric flows in pipes or which are of two-dimensional flows in channels; in any event, note that all flows are three-dimensional in practice, however small the side- and end-effects are. Also Durst, Melling & Whitelaw [3], Cherdron, Durst & Whitelaw [4], Sobey [5], Sobey & Drazin [6], Fearn, Mullin & Cliffe [7], and Durst, Pereira & Tropea [8] describe various recent experimental results of channel flows. They show that although a flow is driven steadily and the cross-section of the channel has a large aspect ratio to simulate two-dimensional flow, the flow may be very unsteady and far from two-dimensional.

It will be convenient to define at the outset the Reynolds number in a form suitable for use throughout the five lectures, namely

$$R = Q/2\nu, \tag{1.1}$$

where Q is the volume flux per unit distance normal to the plane of flow, and ν is the kinematic viscosity of the fluid. Thus we suppose that the flow is driven through the channel with steady flux Q, even though the flow itself may be unsteady, and R acts as a dimensionless scale of the *average* speed of the fluid along the channel. The choice of the factor $\frac{1}{2}$ is not very significant, but it will serve later to simplify the arithmetic a little. This definition of R has the advantage that it does not depend directly on a length scale of the channel, and so is suitable for two-dimensional channels of all shapes and sizes. It also makes it easy for an experimentalist to measure R.

The first theoretical concept of this lecture is as follows. A valuable and widely used framework to understand such flows is to consider the flows in a given channel as they are driven faster, i.e. to consider dynamically similar flows in the channel as the Reynolds number R is slowly increased from zero to infinity. Much can be learnt about a given flow by studying its context, just as we can understand much about a given man's behaviour by studying psychology. Now, Serrin's theorem (cf. the lectures of Professor Galdi, or [9, §53.1]) shows that when R is sufficiently small there is a unique steady flow in the channel, provided that the flow is bounded with given conditions at the inlet and outlet. If the channel is, at least in an idealized thought experiment, two-dimensional then the flow is two-dimensional too. We anticipate that as R is slowly increased a sequence of *bifurcations*, i.e. of qualitative changes of regime of flow, occurs (cf. [9, Chap. 8], [10]), which eventually leads to turbulence for large R. A stable flow becomes unstable as R is increased, and is succeeded by a new stable regime of flow. The sequence begins as the unique (quasi-) steady regime of flow for small values of R becomes unstable when R exceeds some critical value of the Reynolds number, say R_c; the regime is usually replaced by another stable steady flow, but for some channels is replaced by a stable time-periodic flow.

The details of the bifurcations as R increases further are very complicated, and depend on the configuration of the channel. Benjamin [11] proved, by use of Leray–Schauder degree theory, that for a given bounded domain of flow there is a finite positive number of steady solutions for each value of R. Few or none may be stable at a given value of R. Little is known about the unsteady solutions, but we expect to find not only stable steady solutions but also stable periodic, quasi-periodic and chaotic solutions as R increases. At some bifurcation the flow will become strongly three-dimensional, even if the channel simulates a two-dimensional one by having a section of large aspect ratio. If there are multiple equilibria, i.e. many steady flows in the channel are possible at the *same* value of R, then which flow is observed at that value of R will depend on the initial conditions of the flow, so there may be hysteresis such that one stable flow is observed if R is slowly increased from zero to the value and another stable flow if R is slowly decreased from infinity to the value. It is usually found that the spatial as well as the temporal structure of the flow becomes more complicated as R is increased, and eventually the flow becomes turbulent.

The second theoretical concept of this lecture can only be applied to channels with some kind of symmetry. The approach is to exploit the implications of the symmetry. Suppose then that a channel has up-down symmetry. It follows at once that if a flow in the channel is unique then it is symmetric; because for each asymmetric flow there exists another asymmetric flow which is the mirror image of the first flow in the centre plane of the channel. Therefore, at small values of the Reynolds number R there is a unique steady symmetric flow. We have been led to anticipate that as R increases through a critical value R_c the steady symmetric flow will become unstable and there will be a bifurcation either to another steady flow or to a time-periodic flow. This bifurcation must be compatible with the symmetry of the channel. If the *principle of exchange of stabilities* is valid, i.e. if the marginally stable linear mode of the steady flow is non-oscillatory, then the bifurcation will, in general, be to another steady flow; in the present case the steady flow is symmetric Let us examine this in detail. It is helpful to represent the steady flows observed in a channel in a *bifurcation diagram*, i.e. to plot some *state variable*, or variables, describing the flow, against R. The precise choice of state variable, i.e. quantity describing the flow, is not usually important, but it is useful here to choose a variable that vanishes when the flow is symmetric and that in general does not vanish when the flow is asymmetric. So choose the state variable as the transverse, i.e. upward, component, say V, of velocity of the fluid at a given point of the centre plane of the channel. Note that if the flow has up-down symmetry then $V = 0$ (but if $V = 0$ the flow may be asymmetric). Now it can be shown that typically there is either a supercritical (Fig. 1(a)) or subcritical (Fig. 1(b)) pitchfork bifurcation in the bifurcation diagram of the (R, V)-plane. It is a useful and widely used convention to denote a stable solution by a continuous curve and an unstable solution by a dashed curve. So if $0 < R < R_c$ then $V = 0$, i.e. there is a stable symmetric steady flow in the channel, but as R increases through R_c either one of the asymmetric steady solutions with $V \neq 0$ arises continuously (Fig. 1(a)) or the flow To find which substantially different flow arises in the latter case, some non-local analysis not provided by the bifurcation diagram would be necessary. However, the experiments mentioned above all suggest that the former case arises in practice. It is an example of symmetry breaking, and of the Coanda effect whereby a jet clings to one wall rather than flows midway between two walls.

The mathematical theory of a pitchfork bifurcation is elementary and can be found in many textbooks (e.g. [10, §1.5]). Here we merely note that a canonical form to represent a pitchfork bifurcation is the first-order ordinary differential equation

$$\frac{\mathrm{d}V}{\mathrm{d}t} = a(R - R_c)V - lV^3, \tag{1.2}$$

for some state variable V and real constants $a > 0, l$. This is often called a *Landau equation.* The pitchfork is supercritical or subcritical according as the *Landau constant* $l > 0$ or $l < 0$ respectively. You can easily deduce that this gives steady solutions like those depicted in the bifurcation diagrams of Fig. 1 near the point of the R-axis with $R = R_c$. There is also a complex form of this equation for which a or l is complex and the nonlinear term lV^3 is replaced by $l|V|^2V$.

Which of the two asymmetric steady flows arises in the former case of supercritical bifurcation? In the ideal theory the answer would depend upon the initial conditions if the experiment were conducted at a fixed value of the Reynolds number R. If the flow were initially directed upwards near the station where V was measured then the upper steady solution would probably represent the flow when it settled down. In practice, no channel can be made to have the exact up-down symmetry of the theory, the walls being slightly asymmetrical and the inlet not exactly in the middle of the channel, so that the bifurcation diagram would in fact be like either Fig. 2(a) or 2(b). Then the quasi-steady flow would evolve smoothly as R increased slowly. The 'secondary' flow would not be observed unless the experimentalist adopted some other procedure, perhaps by decreasing R from infinity or stirring the flow in a special way at the start of the experiment for a fixed value of R.

If, however, the principle of exchange of stabilities is invalid, i.e. the marginally stable mode of the linear stability of the steady flow is oscillatory, then the typical bifurcation is not a pitchfork bifurcation but is what is called a *Hopf bifurcation*, i.e. the replacement of the symmetric steady flow by a symmetric time-periodic flow, either super- or subcritically as R increases through R_c. In this case the two asymmetric solutions are replaced by one time-periodic solution in each of the bifurcation diagrams of Fig. 1. The time-periodic solution is asymmetric at almost all times, but its mirror image in the centre plane of the channel would be the same periodic solution after a phase change, and so we call it symmetric. Experiments of flow in a symmetric channel have not indicated a Hopf bifurcation, but the linear theory (cf. [9, Chap. 4]) of the instability of plane Poiseuille flow shows that one can occur, at least for some channels.

The breaking of the two-dimensionality (or axisymmetry) of flow in a channel can also be illuminated by such a general approach. Certainly there are *two* symmetries of flows in a channel whose cross-section is a rectangle of large aspect ratio at each station. As the Reynolds number increases, the up-down symmetry (of reflection in the long axis of the rectangle) seems to be broken before the left-right symmetry. To study further symmetry breaking in fluid dynamics, a phenomenon closely related to hydrodynamic stability, the interested reader might start with a recent review [12].

We shall end this lecture by summarizing the theory of instability of a steady flow, and, in particular, of the first instability of the unique steady flow as R increases slowly, i.e. quasi-statically, from zero, in two crucial properties. These properties are for all steady flows, not just those discussed in these lectures. The first property is whether the imaginary part s_i of the eigenvalue s of the least-stable normal mode is zero as the real part s_r increases through zero, i.e. whether the principle of exchange of stabilities is valid. Here we suppose that all velocity components of the normal mode are proportional to e^{st}. (We shall take an example with $s = -\imath kc$ in the next lecture.) This is determined entirely by the linear theory. However, if $s_i = 0$ then in general the steady flow is contiguous to another steady flow at a bifurcation, and if $s_i \neq 0$ then in general the steady flow is contiguous to a time-periodic flow at a Hopf bifurcation. The second crucial property is whether the bifurcation associated with the instability is super- or subcritical. This is determined by the weakly nonlinear theory. If the bifurcation is supercritical then, as R increases slowly, the unique steady flow is succeeded by another stable steady flow or by a stable time-periodic flow according to whether $s_i = 0$ or $s_i \neq 0$ respectively at marginal If the bifurcation is subcritical then, as R increases slowly, the unique steady flow will change abruptly and be succeeded by a flow, steady or unsteady, even turbulent, whose character may be determined only by a strongly nonlinear theory.

Exercises

Q1.1 *The Landau equation.* Show that if V satisfies equation (1.2) and $V(0) = V_0$ then

$$V^2(t) = \frac{a(R - R_c)V_0^2}{lV_0^2 + \{a(R - R_c) - lV_0^2\}\exp\{-2a(R - R_c)t\}},$$

so long as V remains finite. Deduce that the null solution $V(t) = 0$ is stable if $R < R_c$ and unstable if $R > R_c$. Verify also the other results in Fig. 1.

Q1.2 *A Hopf bifurcation.* Consider the equations

$$\frac{\mathrm{d}x}{\mathrm{d}t} = -y + (R - R_c - x^2 - y^2)x, \frac{\mathrm{d}y}{\mathrm{d}t} = x + (R - R_c - x^2 - y^2)y.$$

Show that the only steady solution is the null solution, $x = y = 0$.

Linearizing about this basic solution for small x, y, show that the linearized equations are satisfied by solutions $x, y \propto \mathrm{e}^{st}$, where $s = R - R_c \pm \imath$. Deduce that the null solution is stable if $R < R_c$.

Show that if $R > R_c$ then there exist periodic solutions

$$x(t) = r\cos\theta, y(t) = r\sin\theta$$

of the nonlinear equations, where $r^2 = R - R_c, \theta = t - t_0$ for arbitrary phase t_0.

(Cf. [10, §1.6]).

Q1.3 *Stability of a uniform flow.* Verify that a uniform basic flow of a viscous incompressible fluid with constant velocity $\boldsymbol{U}$ and pressure P gives an exact solution of the Navier–Stokes equations of motion and the equation of continuity. Writing $\boldsymbol{u}(\boldsymbol{x},t) = \boldsymbol{U} + \boldsymbol{u}'(\boldsymbol{x},t)$, and assuming that $\boldsymbol{u}'$ vanishes on the boundary $\partial\Omega$ of the domain Ω of flow (but not that $\boldsymbol{u}'$ is small), deduce that

$$\frac{\mathrm{d}}{\mathrm{d}t}\int_\Omega \frac{1}{2}\boldsymbol{u}'^2 \mathrm{d}\boldsymbol{x} = -\nu \int_\Omega (\partial u_i'/\partial x_j)^2 \mathrm{d}\boldsymbol{x}.$$

Deduce that the flow is stable (in the mean). Discuss the relevance of this result to Serrin's theorem on the stability of a basic flow at sufficiently small values of the Reynolds number.

(Cf. The lectures of Professor Galdi and [9, pp. 27–28, §53.1].)

2. Two Mathematical Models

In this lecture we will describe two asymptotic approaches to the problem of instability of steady flow in a channel and to the weakly nonlinear theory of the corresponding bifurcations. It is desirable to devise a theory that will give us not only some physical insight into the mechanisms of instability but also some numbers which might be satisfactorily related to experimental results.

The first approach is to assume that the walls of the channel are nearly parallel and to approximate the flow in the channel locally by a parallel flow. This is a classic method of the theory of hydrodynamic stability (cf. [9, Chap. 4]) and has been used, e.g., to explain the instability of a boundary layer on a flat plate and of a plane jet. To follow this approach we first consider the stability of an exactly parallel flow.

Accordingly, let us seek to treat the instability of a basic flow through a channel with impermeable parallel walls, at $y = \pm d$ say, assuming that each wall moves in its own plane with prescribed velocity parallel to the x-axis. The Navier–Stokes equations governing two-dimensional flow of a viscous incompressible fluid reduce to the vorticity equation, namely

$$\frac{\partial \zeta}{\partial t} + \frac{\partial(\zeta,\psi)}{\partial(x,y)} = \nu \nabla^2 \zeta, \tag{2.1}$$

where the Cartesian velocity components are

$$u = \frac{\partial \psi}{\partial y} \text{ and } v = -\frac{\partial \psi}{\partial x}, \tag{2.2}$$

the vorticity is

$$\zeta = \frac{\partial v}{\partial x} - \frac{\partial u}{\partial y} = -\nabla^2 \psi, \tag{2.3}$$

and the Laplacian is $\nabla^2 = \partial^2/\partial x^2 + \partial^2/\partial y^2$.

First observe that the mathematical problem is steady, two-dimensional and invariant under the group of translations in the x-direction. Therefore assume that there is a basic flow with the same properties, i.e. that the stream function has the form $\psi = \Psi(y)$. In other words, assume that there is a steady plane parallel flow with velocity $U = \mathrm{d}\Psi/\mathrm{d}y$. Of course, we recognize that there are other flows compatible with the specification of the problem which are unsteady, three-dimensional or not invariant under all translations. Seeking the basic steady parallel flow, we substitute Ψ into the vorticity equation and deduce that

$$\frac{\mathrm{d}^4 \Psi}{\mathrm{d}y^4} = 0. \tag{2.4}$$

This equation and the boundary conditions are a linear problem with a unique solution of the form

$$U(y) = Ay^2 + By + C \tag{2.5}$$

for constants A, B, C determined by the prescribed velocities of the plane walls and the volume flux (or pressure gradient) along the channel. This is *plane Couette–Poiseuille flow.*

To consider the stability of this plane parallel flow, we seek solutions of the form

$$\psi = \Psi + \psi', \tag{2.6}$$

where the perturbation streamfunction is of the form $\psi'(x, y, t)$, linearizing the vorticity equation for small ψ'. The variables x and t are then found to be separable (because the problem is invariant under translation of x or t), so we may consider a complete set of normal modes of the form

$$\psi'(x,y,t) = \Re\{\phi(y)e^{ik(x-ct)}\}. \tag{2.7}$$

The perturbations are thereby shown to be governed by the *Orr–Sommerfeld problem*, which has dimensionless form

$$\phi^{iv} - 2k^2\phi'' + k^4\phi = ikR\{(U-c)(\phi''-k^2\phi) - U''\phi\} \tag{2.8}$$

and

$$\phi, \quad \phi' = 0 \text{ at } y = \pm 1, \tag{2.9}$$

where a prime denotes differentiation with respect to y. This poses some eigenvalue problems. The first is one of temporal modes: to find the set of eigenvalues c and corresponding eigenfunctions ϕ for given U, R and all real wavenumbers k. In this case a mode grows or decays with time like $\exp\{\Im(kc)t\}$ while it oscillates sinusoidally in space and time like $\exp\{i[kx - \Re(kc)t]\}$. It follows that if $\Im(kc) \leq 0$ for all k for all eigenvalues then the basic flow is linearly stable. The principle of exchange of stabilities is said to be valid if $kc = 0$ for the marginally stable mode as the flow becomes unstable, i.e. if not only $\Im(kc)$ increases through zero, but also $\Re(kc)$ varies though zero, as R increases through R_c; then for slightly supercritical R the disturbance grows exponentially where it arises at first and there is said to be ***absolute instability***; in this case there is typically a pitchfork bifurcation, either supercritical or subcritical, as R varies through R_c. Otherwise $\Im(kc) \neq 0$ for the marginally unstable mode and it oscillates and propagates (with its group velocity) as it grows exponentially; there is said to be ***convective instability*** because a small localized disturbance may grow rapidly as it propagates yet decay rapidly where it originated; in this case there is typically a Hopf bifurcation, either supercritical or subcritical, as R varies through R_c. The second problem is one of spatial modes: to find complex k for all real frequencies kc. If $\Im(k) \leq 0$ always then the flow is often said to be stable (provided that the group velocity of the disturbances is positive); this is a useful approach to finding the spatial development of a small disturbance forced by an oscillating source, such as a loudspeaker or a vibrating ribbon, at a point. The general aim is to find the eigenvalue relation between the complex variables c, k and use the relation to solve initial-value problems.

There is also a well developed weakly nonlinear theory of stability of parallel flow. It is based on the asymptotic approximations that the flow is near the margin of stability, i.e. $|R - R_c| \ll 1$, and that the amplitude of the disturbance is small. For the nonlinear theory of the marginally stable mode, we take

$$\psi'(x,y,t) \sim \Re\{A(t)e^{ikx}\phi(y)\}, \tag{2.10}$$

where A is the complex amplitude and k the critical wavenumber of the unique linear mode ϕ which becomes unstable at $R = R_c$. We deduce at length (cf. [9, Chap. 8]) that

$$\frac{\mathrm{d}A}{\mathrm{d}t} = a(R - R_c)A - l|A|^2 A, \tag{2.11}$$

where $a = [\partial(-\imath kc)/\partial R]_{R=R_c}$ and the Landau constant l emerges from a long calculation. We see a pitchfork bifurcation (although it is really a Hopf bifurcation when a or l is complex), supercritical if $\Re(l) > 0$ and subcritical if $\Re(l) < 0$. The invariance of the problem under translation in the x-direction has led to the invariance of the Landau equation under change of phase of A; it is this symmetry of the parallel flow which gives the pitchfork bifurcation, such that the bifurcated flows are waves lacking the translational invariance. If, moreover, the development in space as well as time of a weakly nonlinear marginally stable wavepacket is considered then we add linear dispersion terms and deduce a partial differential equation of the form

$$\frac{\partial A}{\partial t} - b\frac{\partial^2 A}{\partial \xi^2} = a(R - R_c)A - l|A|^2 A, \tag{2.12}$$

where $\xi = x - c_g t$ and c_g is the group velocity of the marginally stable mode. This is a form of the *Ginzburg–Landau equation*,

This is a brief summary of the theory of the Orr–Sommerfeld problem. A lot more about it, in particular three-dimensional modes and the stability characteristics of specific flows, can be found [9, Chap. 4]. Also Eagles and co-authors have applied the theory as an approximation to the stability of flows in a diverging channel with nearly parallel walls [13–15]. Here we shall go on at once to describe the analogous theory of radial flows and their instability. Use the analogy to make the understanding of both problems easier [16].

The problem of the radial basic flow is due independently to Jeffery [17] in 1915 and Hamel [18] in 1916. For this we take plane polar coordinates (r, θ) and consider flow between two impermeable planes with equations $\theta = \pm\alpha$ driven by a steady line source or sink of strength Q at the intersection $r = 0$ of the two planes. Also suppose that the velocities of the planes are either zero or prescribed such that they are radial and inversely proportional to r. See Fig. 3. Now in polar coordinates the velocity components are

$$u_r = \frac{\partial\psi}{r\partial\theta} \text{ and } u_\theta = -\frac{\partial\psi}{\partial r}, \tag{2.13}$$

the vorticity equation becomes

$$\frac{\partial\zeta}{\partial t} + \frac{1}{r}\frac{\partial(\zeta, \psi)}{\partial(r, \theta)} = \nu\nabla^2\zeta, \tag{2.14}$$

where the vorticity $\zeta = -\nabla^2\psi$ and now the Laplacian is $\nabla^2 = \partial^2/\partial r^2 + \partial/r\partial r + \partial^2/r^2\partial\theta^2$.

The equation, but not the boundary conditions, is invariant under the continuous group of rotations. This suggests that we seek basic steady flows such that the streamfunction depends only on θ, i.e. that $\psi = \frac{1}{2}Q\Psi(\theta)$, say. In other words we seek a steady, plane, purely radial, basic flow with velocity $U = \frac{1}{2}Q\mathrm{d}\Psi/r\mathrm{d}\theta$. It follows that

$$\frac{\mathrm{d}^4\Psi}{\mathrm{d}\theta^4} + 4\frac{\mathrm{d}^2\Psi}{\mathrm{d}\theta^2} + 2R\frac{\mathrm{d}\Psi}{\mathrm{d}\theta}\frac{\mathrm{d}^2\Psi}{\mathrm{d}\theta^2} = 0, \tag{2.15}$$

where the Reynolds number is $R = Q/2\nu$ as before. The boundary conditions for stationary walls are that

$$\Psi = \pm 1, \quad \mathrm{d}\Psi/\mathrm{d}\theta = 0 \text{ at } \theta = \pm\alpha. \tag{2.16}$$

This is the *Jeffery–Hamel problem.*

The solution of this problem is not unique, because the ordinary differential equation is nonlinear. In fact there is an *infinity* of solutions Ψ for any given pair of the governing dimensionless parameters R, α. The flows may be well summarized pictorially: the velocity profiles of a few of them are sketched in Fig. 4. Gol'dshtik & Shtern [19] and Gol'dshtik, Hussain & Shtern [20] have also followed Jeffery and Hamel in considering the problem with $\alpha = \pi$ and no walls, i.e. the problem of radial flow due to an isolated line source; this is an elegant problem o So we rush on to consider the stability of Jeffery–Hamel flows, by analogy with the Orr–Sommerfeld problem.

To consider the stability of one of these radial flows, we seek solutions of the form

$$\psi = \frac{1}{2}Q(\Psi + \psi'), \tag{2.17}$$

where the perturbation streamfunction is of the form $\psi'(r,\theta,t)$, linearizing the vorticity equation for small ψ'. Then it is found that the variables r or t, but not both, are separable. So the linear stability problem for normal modes of the form $\psi'(r,\theta,t) = \Re\{\mathrm{e}^{st}\phi(r,\theta)\}$ is reducible to a partial-differential eigenvalue problem with independent variables r, θ. Little progress has been made with this temporal problem to find the eigenvalues s. However, Dean (1934) found that there is also a separable linear ordinary-differential eigenvalue problem for steady *spatial* modes, a mode being of the form

$$\psi'(r,\theta) = \Re\{r^\lambda\phi(\theta)\}. \tag{2.18}$$

It follows that the mode grows or decays in space like $\exp\{\Re(\lambda)\ln r\}$ while it oscillates sinusoidally like $\exp\{\imath\Im(\lambda)\ln r\}$. The problem in fact has the dimensionless form

$$\begin{aligned} &\phi^{iv} + \{\lambda^2 + (\lambda-2)^2\}\phi'' + \lambda^2(\lambda-2)^2\phi \\ &\quad = R\{(\lambda-2)\Psi'(\phi'' + \lambda^2\phi) - 2\Psi''\phi' - \lambda\Psi'''\phi\} \end{aligned} \tag{2.19}$$

and

$$\phi,\ \phi' = 0 \text{ at } \theta = \pm\alpha, \tag{2.20}$$

where a prime here denotes differentiation with respect to θ. Two countably infinite families of eigenvalues λ are found. The real part of λ gives the algebraic rate of growth or decay of the mode as a function of r. A condition of spatial stability is that no disturbance grows down- or upstream, so that $\Re(\lambda) \leq 0$ for all modes of one family and $\Re(\lambda) \geq 0$ for all modes of the other family for given Ψ, R, α. This suggests a criterion for stability to *steady* two-dimensional disturbances, but, of course, tells us nothing about oscillatory disturbances, i.e. about normal modes not governed by the principle of exchange of stabilities. It also tells us nothing about three-dimensional disturbances.

We note that stability of a basic flow conventionally concerns the development in time, not space, of small perturbations. So 'spatial stability' of a steady basic flow to steady perturbations strictly concerns only the properties of all steady flows close to the basic flow, not stability at all. However, there is a close relationship between the developments of perturbations in space and time, as we have seen in our discussion of bifurcation theory, and the relationship encourages the common usage 'spatial stability'; indeed, in general spatial in We will leave the matter there for the present, but consider the Jeffery–Hamel flows and their stability more thoroughly in the exercises and the next lecture.

Exercises

Q2.1 *Absolute and convective stability.* Consider the linear model equation

$$\frac{\partial u}{\partial t} = \sigma u - V\frac{\partial u}{\partial x}$$

for $-\infty < x < \infty$, where σ is real and $V > 0$. Note that $u = U$, where $U(x,t) = 0$ for all x, t gives a solution representing a basic state of rest. Taking normal modes with $u = \Re\{e^{ik(x-ct)}\}$, find the dispersion relation giving the complex velocity c as a function of wavenumber k, and deduce a criterion for stability of each mode.

Show that if $u(x,0) = f(x)$, where f is differentiable and $f(x) \to 0$ as $x \to \pm\infty$, then

$$u(x,t) = e^{\sigma t} f(x - Vt)\,.$$

for $t > 0$. Deduce that the null solution $U = 0$ is absolutely stable, but convectively unstable if $\sigma > 0$.

Q2.2 *Derivation of the Orr–Sommerfeld equation.* Using dimensionless variables, linearize the vorticity equation (2.1) to show that

$$\frac{\partial \zeta'}{\partial t} + U\frac{\partial \zeta'}{\partial x} + \frac{d^2U}{dy^2}\frac{\partial \psi'}{\partial x} = R^{-1}\nabla^2\zeta'. \,.$$

Now, using (2.7), deduce equation (2.8).

Q2.3 *A simple exact solution of the Orr–Sommerfeld equation.* Show that the Navier–Stokes equations are invariant under the continuous group of translations $y \mapsto y + \delta$ for all real δ. Deduce that the basic velocity $U(y)\boldsymbol{i} \mapsto U(y+\delta)\boldsymbol{i} = U(y)\boldsymbol{i} + \delta U'(y)\boldsymbol{i} + O(\delta^2)$ as $\delta \to 0$. Hence or otherwise show that a solution of the Orr–Sommerfeld equation (2.8) is given by $k = 0, \phi. = U'$ for all c, R. Does this given an eigensolution?

Q2.4 *Another simple exact solution of the Orr–Sommerfeld equation.* Show that if $U =$ constant then the general solution of equation (2.8) is

$$\phi(y) = A_1 e^{ky} + A_2 e^{-ky} + A_3 e^{ly} + A_4 e^{-ly},$$

where A_1, A_2, A_3, A_4 are arbitrary constants and $l^2 = k^2 + \imath kR(U - c)$. In fact the solutions $e^{\pm ly + k(x-ct)}$ when $k = 0$ but $kc \neq 0$ correspond to Stokes waves due to an oscillating plane boundary.

Q2.5 *A less-simple exact solution of the Orr–Sommerfeld equation.* Show that if $U(y) = y$ then the general solution of equation (2.8) is

$$\phi(y) = A_1 e^{ky} + A_2 e^{-ky} + \frac{1}{k}\int^y \sinh(k(y-z))\Omega(z)\mathrm{d}z,$$

where

$$\Omega(z) = A_3 \mathrm{Ai}(w) + A_4 \mathrm{Bi}(w),$$

A_1, A_2, A_3, A_4 are arbitrary constants, Ai, Bi are Airy functions, and $w = e^{\imath\pi/6}(kR)^{1/3}(z - c - \imath k/R)$.

Q2.6 *The exact general solution of the Jeffery–Hamel problem.* Show that an integral of equation (2.15) is

$$\Psi''' + 4\Psi' + R(\Psi')^2 = A,$$

where $\Psi' = \mathrm{d}\Psi/\mathrm{d}\theta$ and A is a constant of integration. Deduce that

$$\frac{1}{2}(\Psi'')^2 + 2(\Psi')^2 + \frac{1}{3}R(\Psi')^3 = A\Psi' + B,$$

and thence that the solutions Ψ of the boundary-value problem (2.15), (2.16) may be expressed in terms of Jacobian elliptic functions.

([17, 18], cf. [21].)

Q2.7 *Velocity profiles of types I, II_n and III_n for Stokes flow.* Show that if $R = 0$ for fixed θ then the Jeffery–Hamel problem (2.15), (2.16) has solution

$$\Psi(\theta) = \frac{\sin 2\theta - 2\theta\cos 2\alpha}{\sin 2\alpha - 2\alpha\cos 2\alpha}.$$

Q2.8 *Plane Poiseuille flow.* Show that if $\alpha \to 0$ for fixed θ/α then the Jeffery–Hamel problem (2.15), (2.16) has limiting solution $\Psi(\theta) = 3\theta/2\alpha - \theta^3/2\alpha^3$, i.e. $U(\theta) \sim \frac{3}{2}(1-\theta^2/\alpha^2)/\alpha r$.

Q2.9 *Linearization of the vorticity equation.* Using equation (2.17), linearize equation (2.14) to show that

$$\frac{\partial \zeta'}{\partial t} + \frac{1}{r}\frac{\mathrm{d}\Psi}{\mathrm{d}\theta}\frac{\partial \zeta'}{\partial r} + \frac{2}{r^4}\frac{\mathrm{d}^2\Psi}{\mathrm{d}\theta^2}\frac{\partial \psi'}{\partial \theta} + \frac{1}{r^3}\frac{\mathrm{d}^3\Psi}{\mathrm{d}\theta^3}\frac{\partial \psi'}{\partial r} = R^{-1}\nabla^2\zeta'$$

where $\zeta' = -\nabla^2\psi'$. Now, using (2.16), deduce equation (2.8). Why cannot both variables t, r be separated? Use the energy method to show that the flow is stable when $R = 0$.

Q2.10 *Four exact eigensolutions.* Verify the identity,

$$(\Psi^{iv} + 4\Psi'' + 2R\Psi'\Psi'')' = \Psi^{v} + 4\Psi''' + 2R\Psi'\Psi''' + 2R(\Psi'')^2.$$

Deduce that eigensolutions of the problem (2.19), (2.20) are given by (i) $\lambda = 0, \phi = \Psi'$, (ii) $\lambda = 2, \phi = \Psi'$, (iii) $\lambda = -1, \phi = \cos\theta\Psi'$, and (iv) $\lambda = -1, \phi = \sin\theta\Psi'$, provided that $\Psi'' = 0$ at $\theta = \pm\alpha$.

(Dean [22] gave solutions (i), (ii). In fact the vorticity equation (2.14) is invariant under the group of rotations $r \mapsto r, \theta \mapsto \theta + \delta, \psi \mapsto \psi$ for all real δ, and therefore that $\Psi(\theta+\delta) = \Psi(\theta) + \delta\Psi'(\theta) + O(\delta^2)$ as $\delta \to 0$ must also be a solution of the *equation*: this gives $\psi' = \Psi'$, i.e. $\lambda = 0$, $\phi = \Psi'$. It is also invariant under the translations $x \mapsto x + \delta\cos\beta, y \mapsto y + \delta\sin\beta, \psi \mapsto \psi$, for which $\Psi(\theta - \delta r^{-1}\sin(\theta-\beta)) = \Psi(\theta) - \delta r^{-1}\sin(\theta-\beta)\Psi'(\theta) + O(\delta^2)$ as $\delta \to 0$; therefore $\lambda = -1, \phi = \sin(\theta-\beta)\Psi'(\theta)$.)

Q2.11 *The eigensolutions for Stokes flow.* Show that if $R = 0$ then an *even* eigenfunction of the problem (2.19), (2.20) is

$$\phi(\theta) = \frac{\cos\lambda\theta}{\cos\alpha\lambda} - \frac{\cos(\lambda-2)\theta}{\cos\alpha(\lambda-2)},$$

where λ is a zero of $V(-\lambda)$ and V is defined by

$$V(p) = (p+1)\sin 2\alpha + \sin 2\alpha(p+1).$$

Deduce that $2 - \lambda$ is also an eigenvalue. Find similarly the odd eigenfunctions.

Q2.12 *The pressure gradient in Jeffrey-Hamel flow.* Show that in the Jeffrey-Hamel problem the radial pressure gradient is

$$\frac{\partial p}{\partial r} = \frac{\rho Q^2}{4r^3}\left[\left(\frac{d\Psi}{d\theta}\right)^2 + \frac{1}{R}\frac{d^3\Psi}{d\theta^3}\right]$$

where ρ is the density of the fluid. Deduce that there is an adverse pressure gradient at a wall if $d^2U/d\theta^2 < 0$ there. Does $[d^2U/d\theta^2]_{\theta=\alpha}$ change sign from positive to negative for solutions of type II_1 and II_2 as R increases through R_2?

3. Jeffery–Hamel flows

We began this course by looking at flows in a diverging channel, and were led to consider parallel and radial basic flows and their instability. Parallel flows and their stability are very well known (cf. [9, Chap. 4]), so we shall not give them more time here. However, radial flows are less well understood, and their stability has been treated more recently, so we will review them now. This lecture, then, will re-cover the ground of the second part of the last lecture in detail. First look again at the velocity profiles of some of the flows (Fig. 4).

The Jeffery–Hamel solutions can be found in explicit terms of Jacobian elliptic functions, as Jeffery and Hamel themselves showed. The qualitatively different types of solution have been classified exhaustively and named by Fraenkel [21]. Buitrago [23] has given more detailed properties of them. Yet most of the types will be found to be very unstable and so of little practical significance. The only types which are useful to remember are $I, II_1, II_2, III_1, IV_1, V_1$. However, for the record, we shall briefly list all the types. The flows of types I, II_1 are symmetric in $\pm\theta$ and have uniformly outward velocity, i.e. $u_r \geq 0$ for $-\alpha \leq \theta \leq \alpha$. Flows of type II_n are symmetric and have *net* outflow, such that u_r has $2n-2$ zeros for $-\alpha \leq \theta \leq \alpha$, for $n = 1, 2, \dots$. The flows of type III_1 are symmetric and have uniformly inward velocity; we may represent them by taking either $R < 0$ (for $Q < 0$) or $\alpha < 0$. The flows of type III_n are symmetric and have net inward flow, such that u_r has $2n-2$ zeros for $-|\alpha| \leq \theta \leq |\alpha|$. The flows of type IV_n are asymmetric with net outflow, such that u_r has $2n-1$ zeros for $-\alpha \leq \theta \leq \alpha$ and $u_r < 0$ for θ near α. The flows of type V_n are the mirror images of those of type IV_n in the centre plane $\theta = 0$ of the channel.

Next let us try to understand the complicated nature of the multiple equilibria. Again, a picture (Fig. 5) is helpful. The regions of the (R, α)-plane in which the chief types of solution (i.e. I, II_1, II_2, III_1) exist are indicated.

The coexistence of many solutions makes a diagram like Fig. 5 inadequate on its own. The bifurcation diagrams of Figs. 6(a),(b) are essential supplements. Imagine that each solution is represented by a curve in the four-dimensional $(R, \alpha, \Psi'(0), \Psi''(0))$-space. Then, say, take a section $R = 20$ and project each curve onto the $(\alpha, \Psi'(0))$- and $(\alpha, \Psi''(0))$-planes. This gives Fig. 6. There is nothing very special about taking $R = 20$, but it gives a section as typical as any. Again, we use a continuous curve to denote what we shall soon show plausibly is a stable solution and a dashed curve to denote an unstable solution. Many of the solutions which are unstable have been omitted to simplify the diagrams. Now can you visualize the subcritical pitchfork bifurcation where the stable solution of type II_1 meets the unstable solutions of types II_2, IV_1, V_1? We happen to be looking at the prongs of the pitchfork sideways on in Fig. 6(a), but see the pitchfork more clearly in the projection shown in Fig. 6(b). When the normal velocity gradient of the solution of type II_1 is zero at both walls, the flow may be changed infinitesimally to a flow of type II_2, IV_1 or V_1; see this in Fig. 4 to understand the bifurcation, its eigenfunction (the derivative of the streamfunction, which satisfies the boundary conditions at the bifurcation point) and the mechanism of instability (an infinitesimal rotation of the steady basic flow about the origin). The functions R_2 and α_2 are defined such that the pitchfork bifurcation occurs at $R = R_2(\alpha)$ or, equivalently, $\alpha = \alpha_2(R)$. In Fig. 5 we have fixed R and used α as the parameter; in fact, the diagrams would be qualitatively similar if we fixed a value of α that was moderately small and used R as a parameter (see Fig. 4).

So far we have taken a lot of time to examine the radial flows, but not demonstrated their instability. First note that there are many bifurcations of one steady radial flow to another, whereas a parallel flow does not bifurcate to another parallel flow (it may bifurcate only to a non-parallel flow). In this sense, it is radial flows, not parallel flows, which are better illustrations of the many bifurcations typical of fluid dynamics. The bifurcations tell us at once quite a lot about the instability of the radial flows.

How do we know that the flows are stable and unstable as indicated? Well, when $|R| \ll 1$ *and* there is nearly Stokes flow, Serrin's theorem indicates stability (it is possible that the flux Reynolds number R is small and a Reynolds number based on the maximum velocity of the Jeffery–Hamel flow at a given radius is large, so there is not a unique stable flow when R is small). However, it should be remembered that Serrin's theorem is strictly valid only for bounded flows. In any event, this plausibly suggests that the flows of types III_1, II_1 are stable when R is small. Again, if $|\alpha| \ll 1$ then the walls are nearly parallel, and the flows of types I, II_1, III_1 approximate plane Poiseuille flow, so the conclusion of stability also comes from Orr–Sommerfeld theory, at least if $R < 3848$ (cf. [9, p. 192]). Further, the typical behaviour at a subcritical pitchfork bifurcation is that a stable solution joins with three unstable ones, as shown in Fig. 1(b) (or that all four solutions are unstable). Again, the linearized theory of the spatial development of steady modes introduced last lecture agrees with our interpretation. It is, however, possible that an *oscillatory* instability occurs at a lower value of α than the value α_2 at the pitchfork bifurcation. This certainly does occur when $\alpha = 0$, because then there is plane Poiseuille flow and the Orr–Sommerfeld theory is known to give the onset of instability with a subcritical Hopf bifurcation at $R = 3848$ due to an entirely different mechanism (extraction of energy from the basic flow in thin shear layers, called critical layers, near the walls). So we may imagine another stability boundary in Fig. 4 which starts at $R = 3848, \alpha = 0$ and goes away (we do not yet know whether it goes left or right), such that the basic flow is susceptible to the oscillatory mode of instability if it lies to the right of the boundary. In fact, $\alpha_2(R) \sim 4.712/R$ as $R \to \infty$ along the stability boundary, so that the Orr–Sommerfeld mechanism of instability occurs at a larger value of R than the Jeffery–Hamel mechanism only if $\alpha < \alpha_2(3848) \approx 4.712/3848$

radians = 0.07 degrees. This is a very small angle!

The spatial eigenvalue problem of Dean [22] also is important. To solve it numerical calculations are necessary, but a few general and particular analytic results are known. It is a real problem so each eigenvalue is either real or one of a complex conjugate pair. There are two families of eigenvalues $\lambda_{\pm n}$ for $n = 1, 2, \ldots$ such that $\Re(\lambda_n) \le \Re(\lambda_{n+1}), \Re(\lambda_{-n}) \ge \Re(\lambda_{-n-1})$, and $\Re(\lambda_{\pm n}) \to \pm\infty$ as $n \to +\infty$. The first family is associated with the spatial development of perturbations near $r = \infty$, so that disturbances grow (relative to the basic flow) from infinity as r decreases if $\Re(\lambda_1) < 2$; this then is a criterion for spatial instability. Similarly the other family is associated with growth of disturbances from $r = 0$, and another criterion for instability is that $\Re(\lambda_{-1}) > 0$. If either of these criteria is satisfied it is impossible to set up the Jeffery–Hamel basic flow in practice, because disturbances from real finite boundaries 'near' $r = \infty$ or $r = 0$ will grow inwards and thereby affect the whole flow; thus end effects will be significant however distant the bo In yet other words, Saint Venant's principle is invalid if $\Re(\lambda_1) < 2$ or $\Re(\lambda_{-1}) > 0$, and we expect not to observe a Jeffery–Hamel flow. Even if $\Re(\lambda_1) > 2$ and $\Re(\lambda_{-1}) < 0$, the fact that the decay of a stable mode is algebraic means that it takes a long distance for the influence of a disturbance to become insignificant.

At the pitchfork bifurcation where $R = R_2(\alpha)$, an eigensolution can be seen [22] to be $\lambda = 0, \phi = \Psi'$; Banks, Drazin & Zaturska [24] used this result to analyse the pitchfork bifurcation locally, finding a *spatial* Landau equation of the form

$$r\frac{\mathrm{d}A}{\mathrm{d}r} = a(\alpha - \alpha_2)A - lA^3, \tag{3.1}$$

for $l < 0$, where R is fixed.

When $R = 0$ the eigenvalue problem is the linear one of Stokes flow, and much is known about it [25–27]. Boundary-value problems in a wedge can be solved by use of a Mellin transform. Results for the case $R = 0$ have been used to justify the description above for $R \ge 0$, although the problems for $R > 0$ are not well understood. In particular, this is the basis of our suggesting that Saint Venant's principle is invalid for some steady flows in a diverging channel.

The problem of the linear stability of a basic flow in a wedge demands specification of the boundary conditions either as $r \to 0, \infty$ or at $r = r_1, r_2$ in addition to those at the side walls $\theta = \pm\alpha$ in order for the problem t It is then a partial-differential eigenvalue problem with a simple, but not the simplest, configuration. So numerical methods are necessary to solve it. Essentially no results of this problem are known.

The mechanism of instability of Jeffery–Hamel flows is worthy of a simple physical explanation, but we can offer no better or simpler explanation than the following. The flows corresponding to a subcritical pitchfork bifurcation as R increases for fixed α or as α increases for fixed R can be seen in the arrangement of the first six velocity profiles in Fig. 4. Imagine these flows as replacing the state variable in the bifurcation diagram of Fig. 6, as if they were superposed on Fig. 6. Then you may see that the instability is due to a small rotation of the basic flow itself about the line source.

4. Stability of flow in a diverging channel

In today's lecture we apply the results for Jeffery–Hamel flows to divergent channels, and describe some relevant results of computational fluid dynamics. We shall describe the basic steady flows and their stability at the same time, because the two are so closely connected.

First consider a channel with walls of small curvature, sketched in Fig. 7 with exaggeration of the curvature out of artistic licence. If the walls have equations $y = f(x)$ and $y = -g(x)$ then the angles β and γ between the tangent planes to the walls and the centre plane $y = 0$ are given by $\tan\beta = f'(x)$ and $\tan\gamma = g'(x)$. The flow in the channel is approximated locally at station x as a Jeffery–Hamel flow with semi-angle $\alpha(x) = \frac{1}{2}\{\beta(x)+\gamma(x)\} = \frac{1}{2}\{\arctan f'(x) + \arctan g'(x)\}$ for the value of R determined by the steady flux along the channel and the viscosity of the fluid, at least if there is a suitable Jeffery–Hamel flow for that value of R. This idea of Blasius, Fraenkel and Watson [21, pp. 120–21], [28, p. 407] led Sobey & Drazin [6] to conjecture that the flow in the whole channel is stable if the local Jeffery–Hamel flow is stable everywhere, i.e. if $\alpha_m < \alpha_2(R)$, i.e. if $R < R_2(\alpha_m), = R_c$, say, where we define $\alpha_m = \max_{-\infty<x<\infty} \alpha(x)$. It is a plausible conjecture, and fits quite well their calculations of flows in a channel with fairly small curvature. However, it does not fit well the calculations by Fearn, Mullin & Cliffe [7] of flow in a sudden expansion, which the conjecture would seem to deem unstable at a very low value of R_c (even though the discontinuity of channel width implies infi Recently Sobey & Mullin [29] have implied that the agreement of the conjecture with the numerical results of Sobey & Drazin is fortuitous because their results may be inaccurate on account of their use of upwind rather than central differences. This leaves open some crucial questions, which need careful numerical and experimental investigation. There is also the fundamental difficulty that the Jeffery–Hamel theory gives a *sub*critical pitchfork, whereas laboratory and numerical experiments give a *super*critical pitchfork bifurcation at $R = R_c$. Indeed, there is no stable Jeffery–Hamel flow for $\alpha > \alpha_2$, even though stable steady flows may be calculated for channels with such semi-angles between the walls. The inference is that those stable steady flows in a channel are not approximated by Jeffery–Hamel flows, even if the curvature of the walls is everywhere small.

We noted in Lecture 2 that in the linear temporal stability problem the space variables are not separable. So the stability problem is a partial-differential eigenvalue problem. It is difficult also because the precise boundary conditions at the inlet and outlet are physically important [24] when $\alpha(x) > \alpha_m$ somewhere. We have mentioned two approaches to the stability problem, by assuming that the channel walls are nearly parallel or nearly plane. The approximation of large Reynolds number R is valid for few channels in practice because a steady basic flow usually becomes unstable at a moderate value of R, so the approximation is not very helpful. For all these reasons, a direct numerical solution of the problem is needed, and needed with careful modelling of the inlet and outlet conditions.

To resolve some of these questions raised, we can do no better today than to look at the two-dimensional calculations of the basic flow in a channel by Cliffe & Greenfield [30], Sobey & Drazin [6], Fearn, Mullin & Cliffe ([7], and Dennis *et al.* [Of course, the different shapes of the channels, and the different conditions at the inlet and outlet make it difficult to compare results. The variety of results is inevitably complicated. There have been few attempts to find the stability characteristics of such flows numerically. Shapira, Degani & Weihs [32] took channels with piecewise linear expansion, having angles $\alpha = 10°$ to $90°$ and expansion ratios of 2:1 and 3:1. Their numerical solutions of the two-dimensional linear stability problem gave $R_c \approx 60$, for expansion ratio 3:1 and $R_c \approx 130$ for ratio 2:1, such that R_c decreases as α increases. We may compare these results of computational fluid dynamics to the laboratory (three-dimensional) experiments [2–8]. These experiments confirm the supercriticality of the pitchfork bifurcation, and symmetry breaking at $R = R_c$, and give more information of the regimes of flow at higher values of R, although they were not designed especially to find the bifurcations, and again the channels have many different shapes, expansion ratios and aspect ratios.

5. Conclusions

Now let us look back and review the sequence of instabilities and bifurcations of dynamically similar flows in a diverging channel as the Reynolds number R increases slowly from zero to infinity. We shall take a distant view, not noticing details, important as they may be in practice, and so shall review the transition to turbulence for a general flow.

First, let us summarize the essence of the above results governing the first instability of flow in a divergent channel whose walls are nearly parallel or nearly plane as the Reynolds number R slowly increases. If $\alpha_m < 0.07°$ or thereabouts then $R_c \approx 3800$ and the mechanism of instability is the complicated but well known one, associated with transfer of energy from the basic flow to the disturbance at the critical layer In this case there is a subcritical Hopf bifurcation with an abrupt transition to turbulent spots and thereafter turbulence as R increases above R_c. The case includes that of a converging channel with $\alpha_m < 0$. If, however, $\alpha_m > 0.07°$ or thereabouts then $R_c \approx R_2(\alpha_m)$ and there is a supercritical pitchfork bifurcation for a symmetric channel or an imperfect supercritical pitchfork bifurcation for an asymm The mechanism of instability is a rotation of the basic flow. After this first bifurcation there is a sequence of bifurcations leading to turbulence, but the sequence depends on the shape of the channel and is not well known through theory or experiment.

Secondly, we use an idea that has emerged recently from the work of many applying group theory to hydrodynamic stability of symmetric flows (cf. [33]). Our problem of flow in a diverging channel, which is symmetric about both a horizontal and a vertical plane through the centre line of the channel, is a good illustration of the idea. To be a little more specific, let us suppose also that the cross-section of the channel at each station is a rectangle. Such a problem is governed by partial differential equations and boundary conditions which are invariant under a $\boldsymbol{Z}_2 \times \boldsymbol{Z}_2$ group of transformations, representing reflections in the two planes of symmetry. So the *set* of solutions at each value of the Reynolds number R is invariant under this group, although a given solution may not be because it is asymmetric about one plane, the other, or both.

Now, when the Reynolds number R is small the solution is unique and steady, and therefore possesses all the symmetries of the problem. Further, it is a global attractor in the sense that, whatever the initial conditions, the flow settles down to this steady flow after a sufficiently long time. As R increases quasi-statically, this symmetric steady flow becomes unstable and is replaced by an asymmetric steady flow at a pitchfork bifurcation — the symmetry in the plane cutting the rectangle in its short sides seems to be the one broken first. As R increases further, the symmetry in the other plane is broken (perhaps after the flow becomes time periodic at a Hopf bifurcation). As R increases yet further, time-periodic, quasi-periodic and chaotic unsteady flows, and then flows with spatiotemporal chaos develop. (It is the current fashion to call spatiotemporal chaos complexity.) The precise order of the sequence of the bifurcations introducing these states depends upon the details of each particular channel, but the bifurcations occur more rapidly as R increases, and the flow develops smaller and smaller length scales. Thereafter turbulence occurs, and eventually fully developed turbulence ensues at a large value of R. The hypothesis is that the symmetries are first broken, one by one, as R increases and then 'mended' so that the fully developed turbulence has the $\boldsymbol{Z}_2 \times \boldsymbol{Z}_2$ symmetry of the channel. The regaining of symmetry occurs through the increase in size of attractors in phase space and their consequent collision as R increases (cf. [33]). The turbulent flows on average have coherent structures which may retain both or one of the symmetries of the flow. The experimental evidence to support this scenario is scanty for flows in a channel, but good for some other symmetric flows, for example the flow between coaxial rotating cylinders.

Let us modify this argument by removing the assumption of symmetry and thereby apply it to flow in an asymmetric channel or, indeed, to any class of dynamically similar flows, as the Reynolds number R increases quasi-statically. Again, Serrin's theorem gives a unique global attractor for sufficiently small values of R, and we may make the plausible but crucial assumption that there is another attractor which is unique (i.e. not decomposable into more than one disjoint attractor Again, as R increases, the spatial structure of the flow first increases in complexity in space and time and then decreases in complexity on average when the domains of attraction of the several coexisting attractors grow and collide in phase space as R increases: the essential phenomenon does not depend upon symmetry.

Exercises

Q5.1 *Derivation of the Lorenz system as a model of thermal convection.* You are *given* that the perturbations $\boldsymbol{u}$ and θ of the velocity and temperature fields respectively of a fluid heated from below are governed by the Boussinesq equations, namely

$$\frac{\partial \boldsymbol{u}}{\partial t} + \boldsymbol{u}.\nabla \boldsymbol{u} = -\nabla p + \sigma\theta \boldsymbol{k} + \sigma \nabla^2 \boldsymbol{u},$$

$$\nabla.\boldsymbol{u} = 0,$$

$$\frac{\partial \theta}{\partial t} + \boldsymbol{u}.\nabla\theta = Raw + \nabla^2\theta,$$

for $-\infty < x < \infty, 0 \leq z \leq \pi$, where $\sigma = \nu/\kappa$ is the Prandtl number of the fluid, $Ra = \alpha g d^3 \Delta T/\kappa\nu$ is the Rayleigh number, and dimensionless variables are used with units d/π of length, d^2/κ of time and $\Delta T/R$ of temperature. Here d is the depth of the layer of fluid, ΔT the temperature difference imposed across the layer, α the coefficient of thermal expansion, κ the thermal diffusivity and ν the kinematic viscosity of the fluid. There are 'free' perfectly conducting boundaries at $z = 0, \pi$, so

$$\frac{\partial u}{\partial z} = w = \theta = 0 \quad \text{at} \quad z = 0, \pi.$$

This is the classical model of *Rayleigh–Bénard convection.*

You are further given that there are weakly nonlinear roll cells of the approximate form

$$u(x, z; t) = \sqrt{2}(k^2+1)k^{-1}X(t)S_x C_z,$$

$$w(x, z; t) = -\sqrt{2}(k^2+1)X(t)C_x S_z,$$

$$\theta(x, z; t) = -(k^2+1)^3 k^{-2}\{\sqrt{2}Y(t)C_x S_z + Z(t)S_{2z}\},$$

where $S_x = \sin kx, C_z = \cos z, C_x = \cos kx, S_z = \sin z$ and $S_{2z} = \sin 2z$.

(i) Verify that the equation of continuity is satisfied.

(ii) Verify that the boundary conditions are satisfied.

(iii) Show that the nonlinear terms in the momentum equation are proportional to $\sin 2kx$ or S_{2z}. Hence or otherwise show that the curl of the curl of the momentum equation gives

$$\frac{\mathrm{d}X}{\mathrm{d}\tau} = \sigma(Y - X), \tag{5.1}$$

if appropriate components may be truncated, where $\tau = (k^2 + 1)t$.

(iv) Similarly show that

$$\frac{\mathrm{d}Y}{\mathrm{d}\tau} = rX - Y - ZX, \quad \frac{\mathrm{d}Z}{\mathrm{d}\tau} = -bX + XY, \tag{5.2}$$

where $r = k^2 Ra/(k^2 + 1)^3, b = 4/(k^2 + 1)$. (Hint: you may assume that $\boldsymbol{u}.\nabla\theta = (k^2 + 1)^4(XYS_{2z} + \sqrt{2}ZXC_xS_zC_{2z})/k^2$.)

Q5.2 *Solutions of the Lorenz system.* Use the Lorenz system (5.1)–(5.2) as a model problem of stability, bifurcation, symmetry breaking, onset of chaos and symmetry mending as follows.

Find when the null solution is stable, and explain the physical significance of your findings in terms of Rayleigh–Bénard convection.

Find all the other steady solutions of the Lorenz system. Discuss the physical significance of your findings in terms of Rayleigh–Bénard convection.

Discuss the stability of these other steady solutions and the changes of the set of attractors as r increases from zero to infinity for fixed σ, b. (Hint: you are likely to need to read a book, e.g. [10, §8.1], to do this).

It has been suggested that the Lorenz system models certain features of turbulent convection. Discuss its advantages and disadvantages as a model.

I am grateful to Dr Banks and Dr Zaturska for criticisms of a draft of these notes.

References

[1] Batchelor, G.K. 1967 *An Introduction to Fluid Dynamics.* Cambridge University Press.

[2] Nakayama, Y. ed. 1988 *Visualized Flow.* Oxford: Pergamon Press.

[3] Durst, F., Melling, A. & Whitelaw, J.H. 1974 Low Reynolds number flow over a plane symmetric sudden expansion. *J. Fluid Mech.* **64**, 111–128.

[4] Cherdron, W., Durst, F. & Whitelaw, J.H. 1978 Asymmetric flows and instabilities in symmetric ducts with sudden expansions. *J. Fluid Mech.* **84**, 13–31.

[5] Sobey, I.J. 1985 Observations of waves during oscillatory channel flows. *J. Fluid Mech.* **151**, 395–426.

[6] Sobey, I.J. & Drazin, P.G. 1986 Bifurcations of two-dimensional channel flows. *J. Fluid Mech.* **171**, 263–287.

[7] Fearn, R.M., Mullin, T. & Cliffe, K.A. 1990 Nonlinear flow phenomena in a symmetric sudden expansion. *J. Fluid Mech.* **211**, 595–608.

[8] Durst, F., Pereira, J.C.F. & Tropea, C. 1993 The plane symmetric sudden-expansion at low Reynolds number. *J. Fluid Mech.* **248**, 567–581.

[9] Drazin, P.G. & Reid, W.H. 1981 *Hydrodynamic Stability*. Cambridge University Press.

[10] Drazin, P.G. 1992 *Nonlinear Systems*. Cambridge University Press.

[11] Benjamin, T.B. 1976 Applications of Leray–Schauder degree theory to problems of hydrodynamic stability. *Math. Proc. Camb. Phil. Soc.* **79**, 373–392.

[12] Crawford, J.D. & Knobloch, E. 1991 Symmetry and symmetry-breaking bifurcations in fluid dynamics. *Ann. Rev. Fluid Mech.* **23**, 341–387.

[13] Eagles, P.M. 1966 The stability of a family of Jeffery–Hamel solutions for divergent channel flow. *J. Fluid Mech.* **24**, 191–207.

[14] Allmen, M.J. & Eagles, P.M. 1984 *Stability of divergent channel flows: a numerical approach.* *Proc. R. Soc. Lond.* A **392**, 359–372.

[15] Georgiou, G.A. & Eagles, P.M. 1985 The stability of flows in channels with small wall curvature. *J. Fluid Mech.* **159**, 259–287.

[16] Drazin, P.G. 1988 Perturbations of Jeffery–Hamel flows. On pp. 129–133 of *A Symposium to Honor C.C. Lin*, eds. D.J. Benney, F.H. Shu & C. Yuan. Singapore: World Scientific Press.

[17] Jeffery, G.B. 1915 The two-dimensional steady motion of a viscous fluid. *Phil. Mag.* (6) **29**, 455–465.

[18] Hamel, G. 1916 Spiralförmigen Bewegungen zäher Flüssigkeiten. *Jahresbericht der Deutschen Math. Vereinigung* **25**, 34–60.

[19] Gol'dshtik, M.A. & Shtern, V.N. 1989 Loss of symmetry in viscous flow from a line source. *Fluid Dynamics* **24**, 191–199. Also *Izv. Akad. Nauk SSR Mekh. Zhid. i Gaza* **24**, 35–44.

[20] Gol'dshtik, M.A., Hussain, F. & Shtern, V.N. 1991 Symmetry breaking in vortex-source and Jeffery–Hamel flows. *J. Fluid Mech.* **232**, 521–566.

[21] Fraenkel, L.E. 1962 Laminar flow in symmetrical channels with slightly curved walls. I. On the Jeffery–Hamel solutions for flow between plane walls. *Proc. R. Soc. Lond.* A **267**, 119–138.

[22] Dean, W.R. 1934 Note on the divergent flow of fluid. *Phil. Mag.* (7) **18**, 759–777.

[23] Buitrago, S.E. 1983 Detailed analysis of the higher Jeffery–Hamel solutions. M.Phil. thesis, University of Sussex.

[24] Banks, W.H.H., Drazin, P.G. & Zaturska, M.B. 1988 On perturbations of Jeffery–Hamel flow. *J. Fluid Mech.* **186**, 559–581.

[25] Dean, W.R. & Montagnon, P.E. 1949 On the steady motion of a viscous liquid in a corner. *Proc. Camb. Phil. Soc.* **45**, 389–394.

[26] Sternberg, E. & Koiter, W.T. 1958 The wedge under a concentrated couple: a paradox in the two-dimensional theory of elasticity. *Trans. ASME* E: *J. Appl. Mech.* **25**, 575–581.

[27] Lugt, H.J. & Schwiderski, E.W. 1965 Flows around dihedral angles. I. Eigenmotion and analysis. *Proc. R. Soc. Lond.* A **285**, 382–399.

[28] Fraenkel, L.E. 1963 Laminar flow in symmetrical channels with slightly curved walls. II. An asymptotic series for the stream function. *ProcR. Soc. Lond.* A **272**, 406–428.

[29] Sobey, I.J. & Mullin, T. 1992 Calculation of multiple solutions for the two-dimensional Navier–Stokes equations. *Proc. ICFD Conference, Reading.*

[30] Cliffe, K.A. & Greenfield, A.C. 1982 Some comments on laminar flow in symmetric two-dimensional channels. *Rep.* TP 939. AERE, Harwell.

[31] Dennis, S.C.R., Banks, W.H.H., Drazin, P.G. & Zaturska, M.B. 1994 Flow along a diverging channel. (To be published.)

[32] Shapira, M., Degani, D. & Weihs, D. 1991 Stability and existence of multiple solutions for viscous flow in suddenly enlarged channels. *Computers & Fluids* **18**, 239–258.

[33] King, G.P. & Stewart, I.N. 1992 Symmetric chaos. On pp. 257–315 of *Nonlinear Equations in the Applied Sciences*, eds. W.F. Ames & C.F. Rogers. New York: Academic Press.

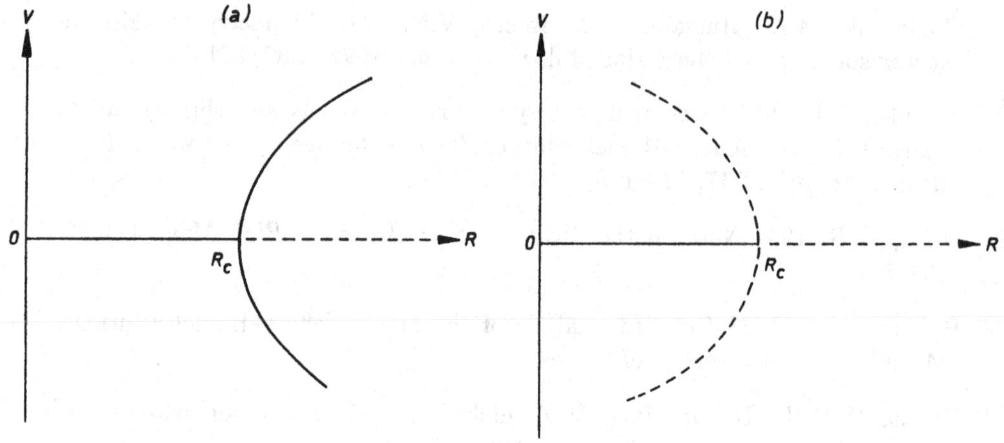

Fig. 1. Pitchfork bifurcations of the Landau equation near the point $(R_c, 0)$ in the (R, V)-plane. (a) Supercritical. (b) Subcritical.

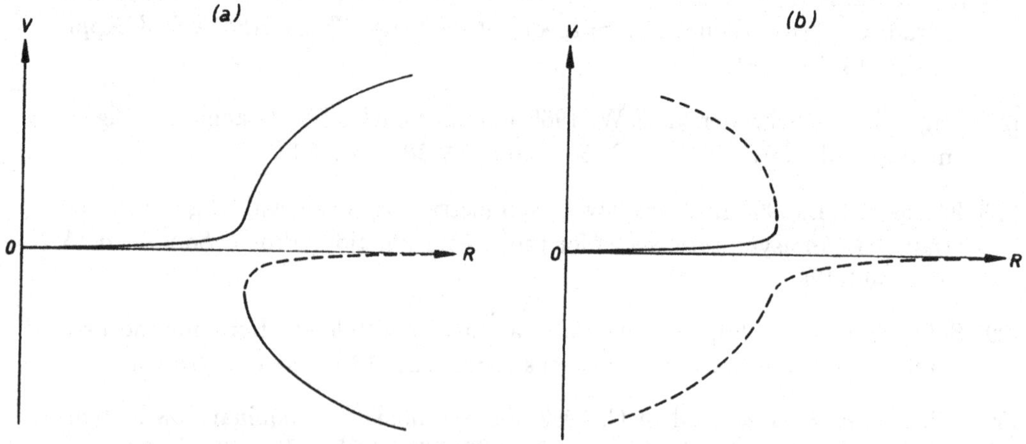

Fig. 2. Imperfect pitchfork bifurcations of the Landau equation in the (R, V)-plane. (a) Supercritical. (b) Subcritical.

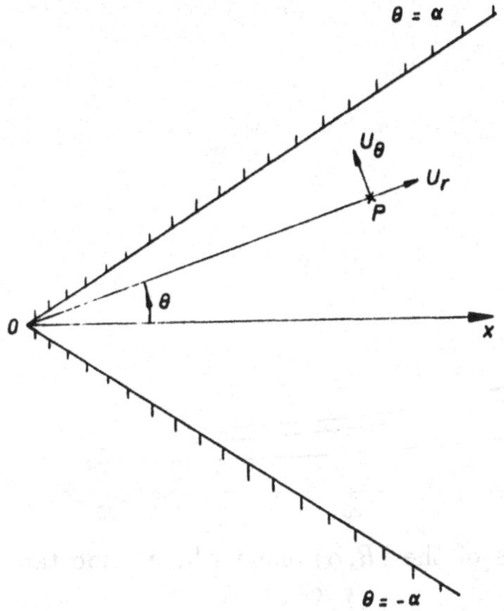

Fig. 3. The configuration of Jeffery–Hamel flows.

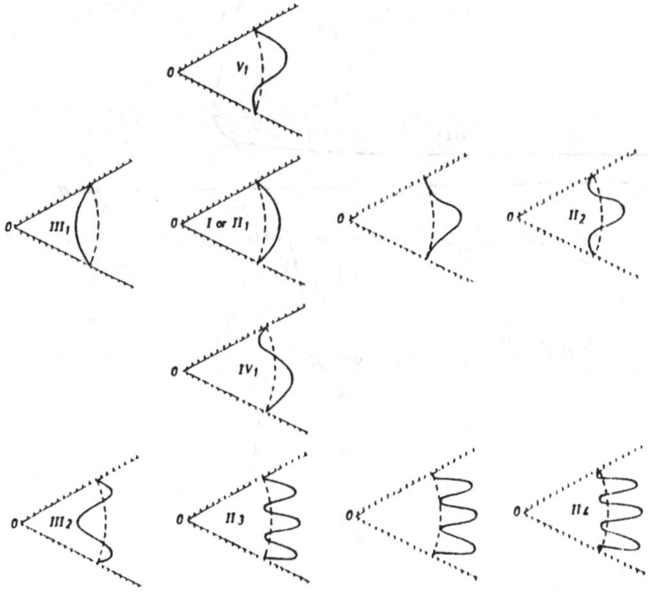

Fig. 4. Some velocity profiles of Jeffery–Hamel flows. The numbers of the types of flow are labelled except for those flows which are at bifurcation and therefore intermediate between one or more types. Note the subcritical pitchfork bifurcation diagram implicit in the arrangement of the upper six diagrams. For practical purposes the six flows are the only important ones.

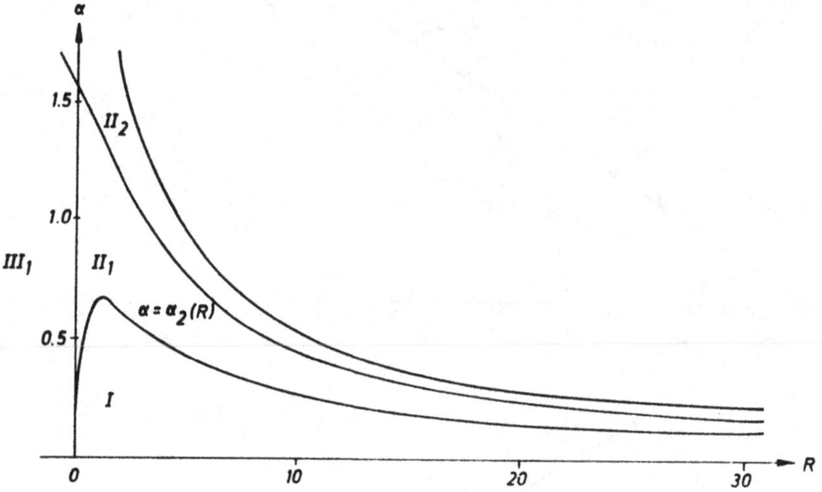

Fig. 5. Boundaries of regions of the (R, α)-plane where important types of Jeffery–Hamel flows exist.

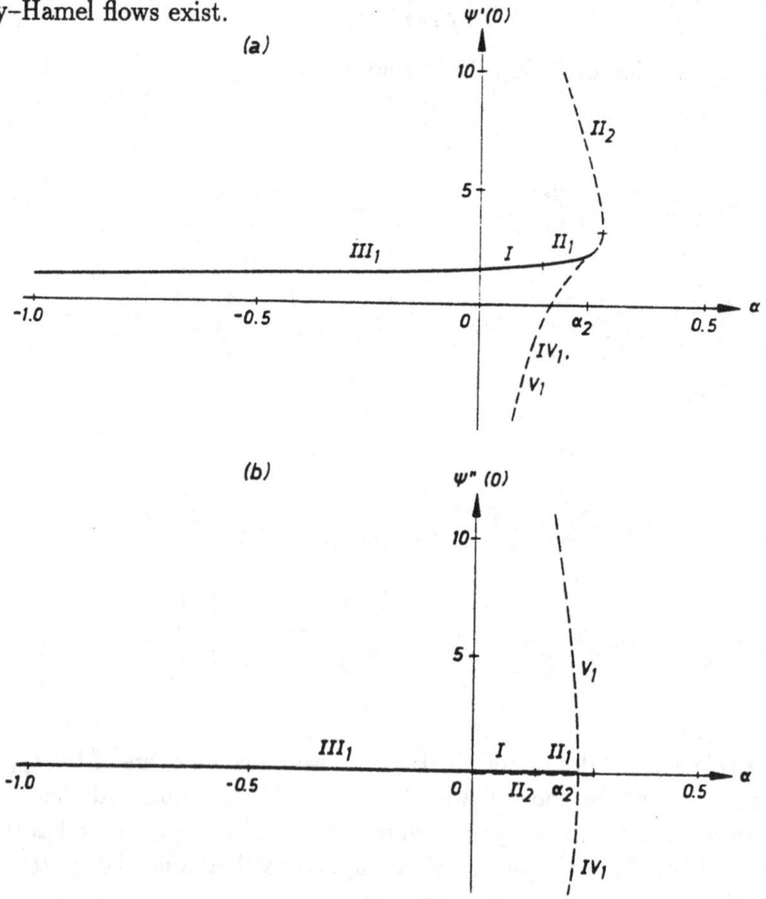

Fig. 6. Bifurcation diagrams of some Jeffery–Hamel flows for $R = 20$. (a) The $(\alpha, \Psi'(0))$-plane. (b) The $(\alpha, \Psi''(0))$-plane.

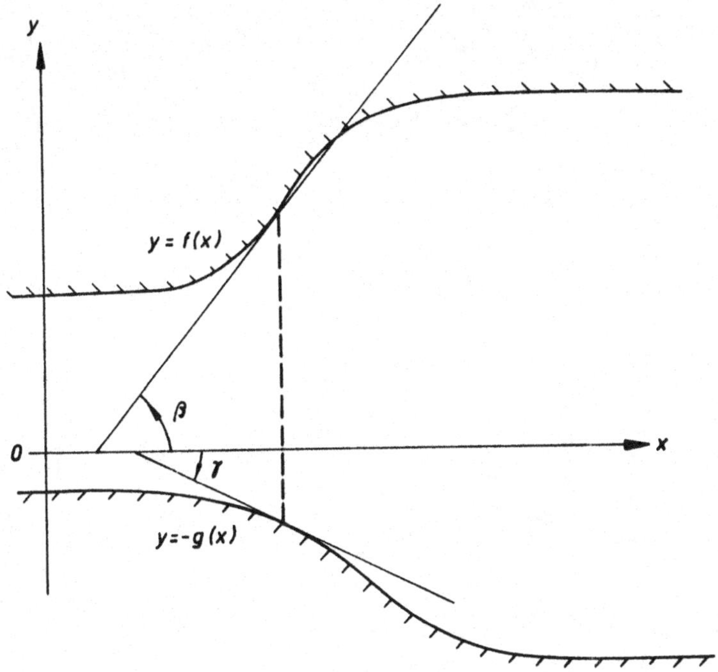

Fig. 7. Sketch of how flow in a channel of small curvature may be locally approximated by Jeffery–Hamel flows.

MATHEMATICAL THEORY
OF SECOND-GRADE FLUIDS

G.P. Galdi
University of Ferrara, Ferrara, Italy

1. Introduction

The aim of these lectures is to study the mathematical properties of the equations governing the motion of a viscous, incompressible second-grade fluid, such as existence, uniqueness of classical solutions and stability of steady-state flows.

As is known, the models describing such fluids have been introduced to explain certain features observed in real fluids which the Newtonian (Navier-Stokes) theory fails to predict. These features include, for instance, *normal stress* in simple shear flows, *shear thickening* and *shear thinning*, *creeping*, etc.[24], [20], [5].

Fluids of second grade are a particular case of a more general class of fluids called *of complexity n* in which the stress tensor $\boldsymbol{T}$ depends on $\boldsymbol{L} = \operatorname{grad} \boldsymbol{v}$ (with $\boldsymbol{v}$ velocity field of the fluid) through the following relation [(1)]

$$\boldsymbol{T} = -\tilde{p}\boldsymbol{I} + \tilde{\boldsymbol{T}}(\boldsymbol{L}, \dot{\boldsymbol{L}}, \ldots, \overset{(n-1)}{\boldsymbol{L}}) \tag{1.1}$$

where $\tilde{p}$ is the pressure field, $n = 1, 2, \ldots$, and $\overset{(n-1)}{\boldsymbol{L}}$ denotes the $(n-1)th$ material derivative of $\boldsymbol{L}$. It is interesting to observe that (1.1) allows the response of the fluid to depend not only on the spatial variation of $\boldsymbol{v}$, as in the Navier-Stokes theory, but also on its (higher) order time derivatives. Thus, if we take $n = 2$, as a special case of (1.1) we obtain the constitutive relation for a second grade fluid (see Dunn and Fosdick [6], Rajagopal [18]):

$$\mathbf{T} = -p\mathbf{I} + \mu\,\mathbf{A}_1 + \tilde{\alpha}_1\,\mathbf{A}_2 + \tilde{\alpha}_2\,\mathbf{A}_1^2 \tag{1.2}$$

(1) We do not take into account the thermal effects.

where μ is the viscosity, $\tilde{\alpha}_1$ and $\tilde{\alpha}_2$ the *normal stress coefficients*, and the tensors $\boldsymbol{A}_1$ and $\boldsymbol{A}_2$ are defined through (Rivlin and Ericksen [20])

$$\boldsymbol{A}_1 = (\mathrm{grad}\,\boldsymbol{v}) + (\mathrm{grad}\,\boldsymbol{v})^T, \tag{1.3}$$

$$\boldsymbol{A}_2 = \frac{d\boldsymbol{A}}{dt} + \boldsymbol{A}_1(\mathrm{grad}\,\boldsymbol{v}) + (\mathrm{grad}\,\boldsymbol{v})^T \boldsymbol{A}_1. \tag{1.4}$$

If a fluid modeled by (1.2) has to be compatible with thermodynamics in the sense that all flows of the fluid meet the Clausius-Duhem inequality, then $\mu \geq 0$. *Throughout this paper we shall assume*

$$\mu > 0.$$

Concerning the normal stress moduli, again from the Clausius-Duhem inequality one shows

$$\tilde{\alpha}_1 + \tilde{\alpha}_2 = 0. \tag{1.5}$$

If, in addition, we make the assumption that the specific Helmholtz free energy is a minimum when the fluid is in equilibrium, then

$$\tilde{\alpha}_1 \geq 0. \tag{1.6}$$

There is a great deal of confusion regarding the status of (1.5), (1.6) and this is discussed in great detail from the physical point of view by Dunn and Rajagopal [7], and we shall not get into a discussion of the same here. However, one of the objectives of these lectures is to investigate to what extent the mathematical properties of the governing equations may change according to whether we assume or not the validity of the above restrictions on the coefficients $\tilde{\alpha}_1$ and $\tilde{\alpha}_2$. For instance, we shall show that condition (1.6) is crucial for stability of steady solutions, while, on the contrary, it does *not* play any role in the existence of these solutions.

Since the fluid is incompressible, it can undergo only isochoric motion, *i.e.*,

$$\mathrm{div}\,\boldsymbol{v} = 0. \tag{1.7}$$

Substituting (1.2) into the balance of linear momentum,

$$\mathrm{div}\,\boldsymbol{T} + \rho\boldsymbol{F} = \rho\frac{d\boldsymbol{v}}{dt},$$

with $\boldsymbol{F}$ body force acting on the fluid, and making use of (1.4), we obtain

$$\begin{aligned} \alpha_1(\Delta\boldsymbol{\omega} \times \boldsymbol{v}) + \alpha_1\frac{\partial\Delta\boldsymbol{v}}{\partial t} + (\alpha_1 + \alpha_2)\left\{\boldsymbol{A}_1\Delta\boldsymbol{v} + 2\mathrm{div}\,\left[(\mathrm{grad}\,\boldsymbol{v})(\mathrm{grad}\,\boldsymbol{v})^T\right]\right\} \\ +\nu\Delta\boldsymbol{v} - \frac{\partial\boldsymbol{v}}{\partial t} - (\boldsymbol{\omega} \times \boldsymbol{v}) = \mathrm{grad}\,P - \boldsymbol{F} \end{aligned} \tag{1.8}$$

where

$$\omega = \operatorname{curl} \boldsymbol{v}, \tag{1.9}$$

and

$$\begin{gathered} P = \frac{1}{\rho}\left\{p - \tilde{\alpha}_1(\boldsymbol{v}\cdot\Delta\boldsymbol{v}) + \frac{\tilde{\alpha}_1}{4}|\boldsymbol{A}_1|^2 + \frac{1}{2}\rho|\boldsymbol{v}|^2\right\} \\ \alpha_1 = \frac{\tilde{\alpha}_1}{\rho} \quad \alpha_2 = \frac{\tilde{\alpha}_2}{\rho} \quad \nu = \frac{\mu}{\rho}. \end{gathered} \tag{1.10}$$

We shall assume throughout that the domain Ω where the motion occurs is a bounded and sufficiently smooth region of the threedimensional space $\mathbb{R}^3$ (2). To (1.7)-(1.10) we append the following side conditions

$$\boldsymbol{v}(x,0) = \boldsymbol{v}_0(x), \quad x \in \Omega \tag{1.11}$$

$$\boldsymbol{v}(y,t) = \boldsymbol{v}_*(y,t), \quad (y,t) \in \partial\Omega \times (0,T) \tag{1.12}$$

where $(0,T)$ denotes the time interval in which we study the motion. In (1.11) $\boldsymbol{v}_0$ denotes a prescribed field representing the velocity distribution of the particle of the fluid at time $t = 0$. Relation (1.12), on the other hand, assumes that the velocity is prescribed also at the boundary. The solenoidality of $\boldsymbol{v}$ in Ω requires

$$\int_{\partial\Omega} \boldsymbol{v}_* \cdot \boldsymbol{n} = 0$$

with $\boldsymbol{n}$ unit outer normal to $\partial\Omega$. It should be remarked that (1.12) is usually assumed for a Navier-Stokes fluid and that it is certainly verified, in this latter case, if the bounding walls are rigid or if they are permeable with $\boldsymbol{v}_*$ denoting the "suction" or "pumping" velocity. However, for the case at hand, it is not obvious, *a priori*, whether (1.12) is still sufficient to ensure well-posedness of problem (1.7)-(1.12) since, unlike the Navier-Stokes theory, the momentum equation contains terms involving *third-order* spatial derivatives. In fact, one can give explicit examples where condition (1.12) alone is not enough to ensure the uniqueness of certain elementary solutions, see Kaloni and Rajagopal [19]. For these solutions to exist it is crucial that the field $\boldsymbol{v}_*$ is not identically zero. In these lectures, for the sake of simplicity, we shall assume the following boundary condition

$$\boldsymbol{v}(y,t) = 0, \quad (y,t) \in \partial\Omega \times (0,T), \tag{1.13}$$

referring the reader to the paper of Galdi, Grobbelaar-Van Dalsen and Sauer [14] for well-posedness of (1.7)-(1.12) in the case $\boldsymbol{v}_* \not\equiv 0$.

(2)The case of an unbounded domain is more difficult to treat and, so far, the only known results are those of Galdi and Rajagopal [16]

We shall also deal with the steady-state counterpart of (1.7)-(1.11), (1.13). This is formally obtained by equating to zero all time derivatives in (1.8) which thus becomes

$$\begin{aligned}\alpha_1(\Delta\omega \times \boldsymbol{v}) + (\alpha_1+\alpha_2)\Big\{\boldsymbol{A}_1\Delta\boldsymbol{v} + 2\operatorname{div}\big[(\operatorname{grad}\boldsymbol{v})(\operatorname{grad}\boldsymbol{v})^T\big]\Big\}\\ +\nu\Delta\boldsymbol{v} - \boldsymbol{\omega}\times\boldsymbol{v} = \operatorname{grad}P - \boldsymbol{F}\end{aligned} \tag{1.14}$$

Of course, in such a case, the body force as well as the velocity and pressure fields depend only on the space variable $x \in \Omega$. To (1.14) we append the boundary condition

$$\boldsymbol{v}(y) = 0, \quad y \in \partial\Omega. \tag{1.15}$$

In these lectures we shall investigate existence and uniqueness of classical solutions to the initial-boundary-value problem (1.7)-(1.11),(1.13) ($\mathcal{IBVP}$) and to the boundary-value problem (1.14)-(1.15) ($\mathcal{BVP}$). Moreover, we shall study the stability of the steady-state solutions. As we mentioned before, to show these results the assumption on the normal stress moduli may play a fundamental role. In fact, the time-dependent properties of solutions will change drastically according to whether condition (1.6) is satisfied or not. In particular, the rest solution to (1.8) (with $\boldsymbol{F}\equiv 0$) is unstable if and only if (1.6) is violated, while if α_1 is positive the $\mathcal{IBVP}$ is well-posed for all times anf for data of suitably restricted size. It is important to emphasize that, for these results to hold, no assumption is needed on the size of $\alpha_1+\alpha_2$. However, the size of $\alpha_1+\alpha_2$ influences the stability region $\mathcal{R}$ of a steady solution and, in particular, $\mathcal{R}\to 0$ as $|\alpha_1+\alpha_2|\to\infty$.

The picture is different for the $\mathcal{BVP}$. In this case, in fact, existence and uniqueness is proved without any assumption on the sign and size of the normal stress moduli. In this way, we have that the thermodynamical restrictions on α_1, α_2 are not required for $\mathcal{BVP}$ being well-posed.

The paper is organized as follows. In Section 2 we introduce some mathematical preliminaries. In Section 3 we study in details suitable transport-like equations which play a fundamental role throughout the paper. In Section 4 we prove existence and uniqueness for $\mathcal{IBVP}$. In particular, we show that if $1/\alpha_1 > -\lambda_1$, with λ_1 first eigenvalue of the Stokes operator, the $\mathcal{IBVP}$ is solvable in a small time interval $[0,T)$ where T depends on the size of the data. Notice that, since $\lambda_1 > 0$, *α_1 is allowed to be also negative.* [(3)] Moreover, if $\alpha_1 > 0$ and the size of the data is suitably restricted, then we can take T arbitrarily large, thus showing existence for all times. However, it should be remarked that to prove such a result we need Ω to be simply connected. If we relax this assumption, we are able to obtain global existence only for α_1 (positive and) sufficiently large, see [13]. In Section 5 we prove existence and uniqueness for $\mathcal{BVP}$. As we already observed, for this study we don't

[(3)] The proof of this result was brought to my attention by Dr. Juha Videman.

need any restriction whatsoever on the normal stress moduli. Finally, in Section 6, we study the nonlinear stability of steady solutions determined in the preceding section. However, also for this study, it is necessary to assume Ω simply connected.

Acknowledgments. Results described in this work are based on the papers [13], [4], [17], [15]. It is my great pleasure to thank all coauthors, namely, V. Coscia, M. Grobbelaar-Van Dalsen, M. Padula, K.R. Rajagopal, N. Sauer and A. Sequeira.

2. Preliminary results.

Throughout the paper we shall need the following notation:

Ω denotes a bounded domaain of $\mathbb{R}^3$.

$H^m(\Omega)$, m a non-negative integer, denotes the Sobolev space $W^{m,2}(\Omega)$ of order m endowed with the norm $\|\cdot\|_m$ and scalar product $(\cdot,\cdot)_m$. With this notation $H^0(\Omega)$ denotes the Hilbert space $L^2(\Omega)$ with norm $\|\cdot\|_0 = \|\cdot\|$, and scalar product $(\cdot,\cdot)_0 = (\cdot,\cdot)$. We recall that, for $m \geq 2$, $H^m(\Omega)$ is an algebra, that is,

$$\|v_1 v_2\|_m \leq c\,\|v_1\|_m\,\|v_2\|_m\,, \quad \text{for all } v_1, v_2 \in H^m(\Omega),$$

with $c = c(\Omega, m)$ (see [1]).

With the class $H^m(\Omega)$ is associated the trace space $H^{m-1/2}(\partial\Omega)$ with norm $\|\cdot\|_{m-1/2,\partial\Omega}$. In similar fashion we denote by $\boldsymbol{H}^m(\Omega)$ and $\boldsymbol{H}^{m-1/2}(\partial\Omega)$ the space of vector functions having components in $H^m(\Omega)$ and $H^{m-1/2}(\partial\Omega)$, respectively. We also set

$$\boldsymbol{V}_m = \boldsymbol{V}_m(\Omega) = \{\boldsymbol{v} \in \boldsymbol{H}^m(\Omega) : \operatorname{div} \boldsymbol{v} = 0\}$$

and

$$\boldsymbol{X}_m = \boldsymbol{X}_m(\Omega) = \{\boldsymbol{v} \in \boldsymbol{V}_m : \boldsymbol{v}\cdot\boldsymbol{n} = 0 \text{ at } \partial\Omega\}$$

where $\boldsymbol{n}$ is the unit outer normal to $\partial\Omega$.

For a given time interval $I = (t_0, t_0 + T)$, $t_0 \geq 0, T > 0$, the open set $\Omega \times I$ will be denoted by Ω_T.

For Y a Banach space with norm $\|\cdot\|_Y$, and $1 \leq p < \infty$ we set

$$L^p(I;Y) = \left\{ v \text{ measurable}, \quad v : t \in I \to v(t) \in Y, \int_0^T \|v(t)\|_Y^p dt < \infty \right\}$$

and denote by $W^{m,p}(I;Y)$ the space of functions such that the distributional time derivatives of order up to and including m are in $L^p(I;Y)$. For $p = \infty$, we denote by $L^\infty(I;Y)$ the Banach space of measurable and essentially bounded functions defined in I with values in Y. The norms in $W^{k,\infty}(I;\boldsymbol{H}^m(\Omega))$ and in $W^{k,\infty}(I;\boldsymbol{H}^{m-1/2}(\Omega))$, $k \geq 0$, are denoted by $\|\cdot\|_{k,m,T}$ and $\|\cdot\|_{k,m-1/2,T,\partial\Omega}$, respectively. For $k = 0$ we write

$\|\cdot\|_{m,T}$ and $\|\cdot\|_{m-1/2,T,\partial\Omega}$. The space of functions of class C^m on I with values in Y is denoted by $C^m(I;Y)$.

Finally, by the symbols c, C we denote generic constants, whose possible dependence on parameters will be specified when necessary. If its numerical value is unessential to our aims, then it may have several different values in a single computation.

We begin to collect some results on certain linear elliptic problems.

Lemma 2.1. *Let Ω be of class C^{m+1}, $m \geq 1$. Then, given*

$$G \in H^{m-1}(\Omega), \quad g \in H^{m-1/2}(\partial\Omega)$$

with

$$\int_{\partial\Omega} g = \int_{\Omega} G,$$

the Neumann problem

$$\Delta p = G \text{ in } \Omega$$

$$\frac{\partial p}{\partial n} = g \text{ at } \partial\Omega$$

admits a unique (up to a constant) solution $p \in H^{m+1}(\Omega)$ such that

$$\|\operatorname{grad} p\|_m \leq c\left(\|G\|_{m-1} + \|g\|_{m-1/2,\partial\Omega}\right).$$

Proof. See, for instance, [21] (4).

This result admits, as a corollary, the following Helmholtz decomposition lemma.

Lemma 2.2. *Let Ω be of class C^{m+1}, $m \geq 0$. Then, given $\boldsymbol{w} \in \boldsymbol{H}^m(\Omega)$ there exist uniquely determined $\boldsymbol{u} \in \boldsymbol{X}_m$, $\operatorname{grad}\tau \in H^m(\Omega)$ such that*

$$\boldsymbol{w} = \boldsymbol{u} + \operatorname{grad}\tau, \quad (\boldsymbol{u}, \operatorname{grad}\tau) = 0.$$

Moreover, the following estimates hold

$$\|\boldsymbol{u}\|_m \leq C\,\|\boldsymbol{w}\|_m\,, \quad \|\operatorname{grad}\tau\|_m \leq C\,\|\boldsymbol{w}\|_m$$

with $C = C(\Omega, m)$.

Proof. Consider the Neumann problem

$$\Delta\tau = \operatorname{div}\boldsymbol{w} \text{ in } \Omega$$

$$\frac{\partial p}{\partial n} = \boldsymbol{w}\cdot\boldsymbol{n} \text{ at } \partial\Omega.$$

(4) The result continues to hold also for $m = 0$ and Ω of class C^2, see the proof of Lemma 2.4.

The result then follows at once from Lemma 2.1.

Remark 2.1. Lemma 2.2 implies the existence of the (orthogonal) projection operator P of $\boldsymbol{L}^2(\Omega)$ into $\boldsymbol{X}_0$. This operator is also linear and continuous from $\boldsymbol{H}^m(\Omega)$ into itself.

In the next lemma we shall show existence and uniqueness for the following Stokes (resolvent) problem:

$$\left.\begin{aligned} \lambda \boldsymbol{v} - \Delta \boldsymbol{v} &= \varphi + \operatorname{grad} \pi \\ \operatorname{div} \boldsymbol{v} &= 0 \end{aligned}\right\} \text{ in } \Omega_T \qquad \boldsymbol{v} = 0 \text{ at } \partial\Omega \times I. \tag{2.1}$$

To this end, denoting by $\mathcal{X}_0^1$ the subspace of $\boldsymbol{H}^1(\Omega)$ constituted by solenoidal functions vanishing at $\partial\Omega$, we set

$$\lambda_1 = \min_{\boldsymbol{v} \in \mathcal{X}_0^1} \frac{\|\operatorname{grad} \boldsymbol{v}\|_0}{\|\boldsymbol{v}\|_0}.$$

As is known, λ_1 is positive and represents the first eigenvalue of the Stokes operator. We have

Lemma 2.3. *Let Ω be of class C^{m+2}, $m \geq 0$. For any $\varphi \in L^\infty(I; \boldsymbol{H}^m)$, $m \geq 0$, and $\lambda > -\lambda_1$, problem (2.1) has a unique solution*

$$(\boldsymbol{v}, \pi) \in L^\infty(I; \boldsymbol{X}_{m+2}) \times L^\infty(I; H^{m+1}(\Omega)).$$

Moreover, there exists a positive constant $C = C(\Omega, m, \lambda)$ such that

$$\|\boldsymbol{v}\|_{m+2,T} + \|\operatorname{grad} \pi\|_{m,T} \leq C\, \|\varphi\|_{m,T}\,.$$

Finally, if $\varphi \in W^{k,\infty}(I; X_m)$, $k \geq 0$, then

$$(\boldsymbol{v}, \pi) \in W^{k,\infty}(I; X_{m+2}) \times L^\infty(I; H^{m+1}(\Omega))$$

and the following estimate holds

$$\|\boldsymbol{v}\|_{k,m+2,T} + \|\operatorname{grad} \pi\|_{k,m,T} \leq C\, \|\varphi\|_{k,m,T}\,.$$

Proof. Using Poincaré's inequality

$$\|\boldsymbol{v}\| \leq \lambda_1 \|\operatorname{grad} \boldsymbol{v}\|$$

and the Riesz representation theorem, by the assumption on λ we can for almost all fixed $t \in \overline{I}$ immediately establish existence of a weak solution satisfying the identity

$$\lambda \|\boldsymbol{v}\|^2 + \|\mathrm{grad}\, \boldsymbol{v}\|^2 = (\boldsymbol{\varphi}, \boldsymbol{v}).$$

Then, by the Schwarz inequality, we deduce

$$\|\boldsymbol{v}\| \leq \frac{1}{\lambda + \lambda_1} \|\boldsymbol{\varphi}\| .$$

This estimate, together with the classical regularity results for the the Stokes problem [3], [2] yields the validity of the first part of the lemma. In a completely analogous way, differentiating (2.1) k times with respect to t, we show the second part.

We have also

Lemma 2.4. *Let Ω be a simply connected domain of class C^{m+3}, $m \geq 0$. Then, for any $\lambda \geq -\lambda_1$ and $\boldsymbol{u} \in W^{k,\infty}(I; \boldsymbol{V}_m)$, $k \geq 0$, there exists one and only one $\boldsymbol{v} \in W^{k,\infty}(I; \boldsymbol{X}_{m+3}(\Omega))$ solving the problem*

$$\left.\begin{array}{l} \boldsymbol{u} = \mathrm{curl}\,(\lambda \boldsymbol{v} - \Delta \boldsymbol{v}) \\ \mathrm{div}\, \boldsymbol{v} = 0 \end{array}\right\} \text{ in } \Omega_T$$

$$\boldsymbol{v} = 0 \text{ at } \partial\Omega \times I.$$

This solution satisfies the estimate:

$$\|\boldsymbol{v}\|_{k,m+3,T} \leq c \|\boldsymbol{u}\|_{k,m,T} \tag{2.2}$$

with $c = c(\Omega, m, \lambda)$.

Proof. It is enough to show the result in the case $m = 0$, since the proof in the general case $m \geq 0$ follows from a simple inductive argument [9] Proposition 1.4, p. 41 and Remarque 1.6. We begin to prove that every $\boldsymbol{u} \in \boldsymbol{V}_0$ can be written as

$$\boldsymbol{u} = \mathrm{curl}\, \boldsymbol{f}, \quad \boldsymbol{f} \in \boldsymbol{X}_1 \tag{2.3$_1$}$$

with

$$\|\boldsymbol{f}\|_1 \leq c \|\boldsymbol{u}\|. \tag{2.3$_2$}$$

It is clear that $\boldsymbol{f}$ is uniquely determined since, if $\boldsymbol{f}_1$ satisfies the same property as $\boldsymbol{f}$, we would have $\mathrm{curl}\,(\boldsymbol{f} - \boldsymbol{f}_1) = 0$, that is, $\boldsymbol{f} - \boldsymbol{f}_1 = \mathrm{grad}\, \psi$. Being $(\boldsymbol{f} - \boldsymbol{f}_1) \in \boldsymbol{X}_1$ it follows that ψ satisfies a homogeneous Neumann problem in Ω leading to $\psi = \mathrm{const}$.

We make a suitable extension of the field $\boldsymbol{u}$ as follows. Consider the Neumann problem:

$$\begin{aligned} &\Delta p = 0 \ \text{ in } \Omega_R \\ &\frac{\partial p}{\partial n} = \boldsymbol{u}\cdot\boldsymbol{n} \ \text{ at } \partial\Omega \\ &\frac{\partial p}{\partial n} = 0 \ \text{ at } \partial B_R \end{aligned} \tag{2.4}$$

where B_R is an open ball containing $\overline{\Omega}$ and $\Omega_R = B_R - \Omega$. Since

$$\int_{\partial\Omega} \boldsymbol{u}\cdot\boldsymbol{n} = 0,$$

it is well known that problem (2.4) admits only one solution (in the distributional sense) with $\int_\Omega p = 0$ and such that:

$$\|p\|_1 \le c\|\boldsymbol{u}\cdot\boldsymbol{n}\|_{-1/2,\partial\Omega}. \tag{2.5}$$

We shall briefly sketch the proof. Since $\boldsymbol{u} \in \boldsymbol{V}_0$, from well-known results on the trace, we have $\boldsymbol{u}\cdot\boldsymbol{n} \in H^{-1/2}(\partial\Omega)$. (5) We can then put problem (2.4) in the following weak form

$$\int_\Omega \operatorname{grad} p \cdot \nabla\psi = < \boldsymbol{u}\cdot\boldsymbol{n}, \psi > \tag{2.6}$$

for all $\psi \in \widetilde{H}^1(\Omega) := \{\psi \in H^1(\Omega) : \int_\Omega \psi = 0\}$ and where $< \cdot,\cdot >$ is the duality pairing between $H^{-1/2}(\partial\Omega)$ and $H^{1/2}(\partial\Omega)$. By Riesz theorem, one easily shows that (2.6) admits at least one solution $p \in \widetilde{H}^1(\Omega)$. Furthermore, choosing $\psi = p$ in (2.6) and using the trace inequality (see [1])

$$\|p\|_{1/2,\partial\Omega} \le c\|\operatorname{grad} p\|, \qquad p \in \widetilde{H}^1(\Omega),$$

we find:

$$\|\operatorname{grad} p\|^2 \le c\|\boldsymbol{u}\cdot\boldsymbol{n}\|_{-1/2,\partial\Omega}\|\operatorname{grad} p\|$$

which, in turn, implies (2.5). We next define the following extension of the field $\boldsymbol{u}$:

$$\tilde{\boldsymbol{u}} = \begin{cases} \boldsymbol{u} & \text{in } \Omega \\ \operatorname{grad} p & \text{in } \Omega_R \end{cases}$$

It is clear from (2.4) that $\operatorname{div}\tilde{\boldsymbol{u}} = 0$ (in the sense of distributions) and, moreover, $\tilde{\boldsymbol{u}}\cdot\boldsymbol{n} = 0$ at ∂B_R. We now set

$$\tilde{\boldsymbol{f}} = -\operatorname{curl}(\mathcal{E} * \tilde{\boldsymbol{u}}), \tag{2.7}$$

(5) We recall that, by well-known trace theorems (see, e.g., [11]) the normal component $\boldsymbol{u}\cdot\boldsymbol{n}$ of $\boldsymbol{u} \in \boldsymbol{V}_0$ at the boundary $\partial\Omega$ of Ω belongs to $H^{-1/2}(\partial\Omega)$, the dual space of $H^{1/2}(\partial\Omega)$.

where $\mathcal{E}$ is the Laplace fundamental solution and $*$ means convolution. Since $\tilde{\boldsymbol{u}}$ is divergence free, it follows that

$$\operatorname{curl} \tilde{\boldsymbol{f}} = -\operatorname{curl}\operatorname{curl}(\mathcal{E} * \tilde{\boldsymbol{u}}) = \Delta(\mathcal{E} * \tilde{\boldsymbol{u}}) = \tilde{\boldsymbol{u}}.$$

In addition, from (2.7) and from the Calderón-Zygmund theorem (*cf.* e.g., [11]) we find

$$\|\tilde{\boldsymbol{f}}\|_1 \leq c\|\boldsymbol{u}\|.$$

We define

$$\boldsymbol{f} = \tilde{\boldsymbol{f}} - \operatorname{grad}\psi$$

where

$$\Delta\psi = 0 \ \text{ in } \Omega$$
$$\frac{\partial\psi}{\partial n} = \boldsymbol{f}\cdot\boldsymbol{n} \ \text{ at } \partial\Omega.$$

It is easily seen that the field $\boldsymbol{f}$ satisfies conditions $(2.3_{1,2})$. We now determine a unique solution (see Lemma 2.3) to the problem:

$$\left.\begin{aligned} \lambda\boldsymbol{v} - \Delta\boldsymbol{v} &= \boldsymbol{f} + \operatorname{grad}\pi \\ \operatorname{div}\boldsymbol{v} &= 0 \end{aligned}\right\} \text{ in } \Omega \qquad (2.8)$$
$$\boldsymbol{v} = 0 \ \text{ at } \partial\Omega$$

which satisfies the estimate:

$$\|\boldsymbol{v}\|_3 \leq c\|\boldsymbol{f}\|_1,$$

and so (2.2) is proved. To show uniqueness, it is sufficient to observe that if $\operatorname{curl}(\lambda\boldsymbol{v} - \Delta\boldsymbol{v}) = 0$, by the hypothesis of simple connectedness on Ω, we derive

$$\lambda\boldsymbol{v} - \Delta\boldsymbol{v} = \operatorname{grad}\pi$$

for some $\pi \in H^2(\Omega)$. Since $\operatorname{div}\boldsymbol{v} = 0$ in Ω and $\boldsymbol{v} = 0$ at $\partial\Omega$, it follows $\boldsymbol{v} = 0$, $\pi = \text{const}$. Finally, differentiating (2.3_2) and (2.8) k times with respect to t and using uniqueness, one readily finds that $\boldsymbol{v} \in W^{k,\infty}(I;\boldsymbol{X}_1)$ completing the proof of the lemma.

We conclude this section with a result on a differential inequality

Lemma 2.5. *Let $y(t)$ be a smooth positive function in $[0,T]$ satisfying the inequality*

$$y'(t) + (k_1 - k_2 y^{\rho}(t))\, y(t) \leq F(t), \quad \text{for all } t \in [0,T], \qquad (2.9)$$

where $k_1 > 0$, $k_2 \in \mathbf{R}$, $\rho \geq 0$ and

$$\int_0^T F(t)dt < \infty.$$

Moreover, let $\varepsilon > 0$ be such that $k_1 - k_2\,\varepsilon^\rho = k > 0$. Then, if

$$\int_0^T F(t)\,dt < \frac{\varepsilon}{2}, \qquad y(0) < \frac{\varepsilon}{2}$$

it follows that $y(t) < \varepsilon$, for all $t \in [0,T]$.

Proof. Assume by contradiction that, for some $\bar{t}$ we have

$$y(\bar{t}) = \varepsilon \quad \text{and} \quad y(t) < \varepsilon, \;\; \forall t \in [0,\bar{t}) \,.$$

Since $k_1 - k_2\,\varepsilon^\rho = k > 0$ we obtain the inequality

$$y'(t) + k\,y(t) \le F(t)\,, \qquad \forall t \in [0,\bar{t}]\,,$$

which integrated in $[0,\bar{t}\,]$ gives in particular

$$y(\bar{t}) \le y(0) + \int_0^{\bar{t}} F(s)\,ds\,,$$

hence

$$y(\bar{t}) \le y(0) + \delta < \frac{\varepsilon}{2} + \delta < \varepsilon$$

for δ small enough.

3. Existence and Uniqueness of Solutions to Certain Transport Equations.

The objective of this section is to investigate existence, uniqueness and corresponding estimates for certain transport-like equations. Specifically, on the one hand, we shall study the initial-boundary-value problem:

$$\left.\begin{aligned} &\frac{\partial \boldsymbol{u}}{\partial t} + \sigma(\boldsymbol{u} - \boldsymbol{v}) + \operatorname{curl}\boldsymbol{u} \times \boldsymbol{v} = -\operatorname{grad} p + \boldsymbol{N}(\boldsymbol{v}) + \boldsymbol{F} \\ &\operatorname{div}\boldsymbol{u} = 0 \end{aligned}\right\} \text{ in } \Omega_T \tag{3.1}$$

$$\boldsymbol{u}\cdot\boldsymbol{n} = 0 \;\text{ at } \partial\Omega \times (t_0, t_0 + T)$$

$$\boldsymbol{u}(\cdot, t_0) = \boldsymbol{u}(t_0), \;\text{ in } \Omega,$$

where $\sigma = \nu/\alpha_1$, $\boldsymbol{F}$ and $\boldsymbol{u}(t_0)$ are prescribed vector functions and

$$\boldsymbol{N}(\boldsymbol{v}) := -(\alpha_1 + \alpha_2)\left\{\boldsymbol{A}_1 \Delta \boldsymbol{v} + 2\operatorname{div}\left[(\operatorname{grad}\boldsymbol{v})(\operatorname{grad}\boldsymbol{v})^T\right]\right\}. \tag{3.2}$$

On the other hand, we shall investigate the same question for the following boundary-value problem

$$\left.\begin{aligned} &\nu \boldsymbol{u} + \alpha_1 \operatorname{curl} \boldsymbol{u} \times \boldsymbol{v} - \boldsymbol{\omega} \times \boldsymbol{v} = -\operatorname{grad} p + \boldsymbol{N}(\boldsymbol{v}) + \boldsymbol{F} \\ &\operatorname{div} \boldsymbol{u} = 0 \end{aligned}\right\} \text{ in } \Omega \tag{3.3}$$

$$\boldsymbol{u} \cdot \boldsymbol{n} = 0 \ \text{ at } \partial\Omega$$

where $\boldsymbol{\omega}$ is given in (1.9).

We begin to prove some properties of the operator $\boldsymbol{N}(\boldsymbol{v})$ which are easily established by means of the Sobolev embedding theorems and by recalling that, for $m \geq 2$, $H^m(\Omega)$ is an algebra.

Lemma 3.1. *Let $m \geq 1$. Then the following estimate holds*

$$\|\boldsymbol{N}(\boldsymbol{v}_1) - \boldsymbol{N}(\boldsymbol{v}_2)\|_m \leq c(\|\boldsymbol{v}_1\|_{m+2} + \|\boldsymbol{v}_2\|_{m+2})\|\boldsymbol{v}_1 - \boldsymbol{v}_2\|_{m+2},$$

for all $\boldsymbol{v}_1, \boldsymbol{v}_2 \in \boldsymbol{H}^{m+2}(\Omega)$. Thus, in particular,

$$\|\boldsymbol{N}(\boldsymbol{v})\|_m \leq c\|\boldsymbol{v}\|_{m+2}.$$

In these inequalities, $c = c(\Omega, m, |\alpha_1 + \alpha_2|)$.

Proof. By the definition of $\boldsymbol{N}$ we see that, setting $\boldsymbol{w} = \boldsymbol{v}_1 - \boldsymbol{v}_2$, all terms in the expression $\boldsymbol{N}(\boldsymbol{v}_1) - \boldsymbol{N}(\boldsymbol{v}_2)$ have the form

$$D(\boldsymbol{v}_1 + \boldsymbol{w})D^2(\boldsymbol{v}_1 + \boldsymbol{w}) - D(\boldsymbol{v}_1)D^2(\boldsymbol{w}) = D\boldsymbol{w}D^2\boldsymbol{v} + D\boldsymbol{v}D^2\boldsymbol{w} + D\boldsymbol{w}D^2\boldsymbol{w},$$

where D^k is a differential operator of order k. Consider first the case $m = 1$. We have

$$\|D\boldsymbol{w}D^2\boldsymbol{v}\|_1 \leq \|D\boldsymbol{w}D^2\boldsymbol{v}\| + \|D^2\boldsymbol{w}D^2\boldsymbol{v}\| + \|D\boldsymbol{w}D^3\boldsymbol{v}\|$$

and so, by the Sobolev inequalities [1],

$$\|Df\|_{L^\infty(\Omega)} \leq c\|f\|_3, \quad \|D^2 f\|_{L^4(\Omega)} \leq c\|f\|_3$$

and the Schwarz inequality we find

$$\|D\boldsymbol{w}D^2\boldsymbol{v}\|_1 \leq c\|\boldsymbol{v}\|_3\|\boldsymbol{w}\|_3.$$

If $m \geq 2$ we use the algebra property of $H^m(\Omega)$ to deduce

$$\|D\boldsymbol{w}D^2\boldsymbol{v}\|_m \leq \|D\boldsymbol{w}\|_m\|D^2\boldsymbol{v}\|_m \leq c\|\boldsymbol{v}\|_{m+2}\|\boldsymbol{w}\|_{m+2}.$$

Since the same reasoning applies to the other terms, the lemma is proved.

In the next lemma we establish existence and uniqueness for (3.1).

Lemma 3.2. *Assume that* $m \geq 2$,

$$\boldsymbol{v} \in L^\infty(I, \boldsymbol{X}_{m+2}), \quad \textit{with } \|\boldsymbol{v}\|_{m+2,T} \leq M, \quad \boldsymbol{F} \in L^\infty(I, \boldsymbol{H}^m(\Omega)),$$

and that $\boldsymbol{u}(t_0) \in \boldsymbol{X}_m$. *Then, there exists a unique solution* $\boldsymbol{u}, \operatorname{grad} p$ *to* (3.1) *such that*

$$\boldsymbol{u} \in L^\infty(I; \boldsymbol{X}_m) \cap W^{1,\infty}(I; \boldsymbol{X}_{m-1}), \quad \operatorname{grad} p \in L^\infty(I; \boldsymbol{H}^m(\Omega))$$

$$\|\boldsymbol{u}\|_{m,T} + \left\| \frac{d\boldsymbol{u}}{dt} \right\|_{m-1,T} \leq C,$$

with $C = C(\Omega, m, M, T, \nu, \alpha_1, \alpha_2, \|\boldsymbol{u}(t_0)\|_m, \|\boldsymbol{F}\|_{m,T})$.

Proof. We begin to derive an *a priori* estimate for the solutions to (3.1). To this end, we notice that in view of the identity

$$\operatorname{curl} \boldsymbol{u} \times \boldsymbol{v} = -\operatorname{grad}(\boldsymbol{u} \cdot \boldsymbol{v}) + \boldsymbol{v} \cdot \operatorname{grad} \boldsymbol{u} + \boldsymbol{u} \cdot \operatorname{grad} \boldsymbol{v} + \boldsymbol{u} \times \operatorname{curl} \boldsymbol{v}, \tag{3.4}$$

equation (3.1_1) can be put in the following equivalent form

$$\frac{\partial \boldsymbol{u}}{\partial t} + \sigma(\boldsymbol{u} - \boldsymbol{v}) + \boldsymbol{v} \cdot \operatorname{grad} \boldsymbol{u} = -\boldsymbol{u} \cdot \operatorname{grad} \boldsymbol{v} - \boldsymbol{u} \times \operatorname{curl} \boldsymbol{v} - \operatorname{grad} \tau + \boldsymbol{N}(\boldsymbol{v}) + \boldsymbol{F} \tag{3.5}$$

where

$$\tau = p - \boldsymbol{u} \cdot \boldsymbol{v}.$$

We apply the derivative operator D^k (k is a multi-index) to both sides of (3.5), take the scalar product in $L^2(\Omega)$ with $D^k\boldsymbol{u}$ and sum over k, with $0 \leq |k| \leq m$. We thus obtain

$$\begin{aligned} \tfrac{1}{2}\frac{d}{dt}\|\boldsymbol{u}\|_m^2 + \sigma\|\boldsymbol{u}\|_m^2 = \sigma(\boldsymbol{v}, \boldsymbol{u})_m &- (\boldsymbol{v} \cdot \operatorname{grad} \boldsymbol{u}, \boldsymbol{u})_m - (\boldsymbol{u} \cdot \operatorname{grad} \boldsymbol{v}, \boldsymbol{u})_m \\ &-(\boldsymbol{u} \times \operatorname{curl} \boldsymbol{v}, \boldsymbol{u})_m - (\operatorname{grad} \tau, \boldsymbol{u})_m \\ &+(\boldsymbol{N}(\boldsymbol{v}), \boldsymbol{u})_m + (\boldsymbol{F}, \boldsymbol{u})_m. \end{aligned} \tag{3.6}$$

Let us now give appropriate bounds for the right-hand side of (3.6). Using the Schwarz inequality and the algebra property of the space $H^m(\Omega)$, $m \geq 2$, we deduce

$$|(\boldsymbol{u} \cdot \operatorname{grad} \boldsymbol{v}, \boldsymbol{u})_m| + |(\boldsymbol{u} \times \operatorname{curl} \boldsymbol{v}, \boldsymbol{u})_m| \leq c\|\boldsymbol{v}\|_{m+2}\|\boldsymbol{u}\|_m^2. \tag{3.7}$$

Moreover, again by the Schwarz inequality and Lemma 3.1, it follows that

$$\begin{aligned} |(\boldsymbol{v}, \boldsymbol{u})_m| &+ |(\operatorname{grad} \tau, \boldsymbol{u})_m| + |(\boldsymbol{N}(\boldsymbol{v}), \boldsymbol{u})_m| \\ &+ |(\boldsymbol{F}, \boldsymbol{u})_m| \leq c(\|\boldsymbol{v}\|_{m+2} + \|\boldsymbol{F}\|_m + \|\operatorname{grad} \tau\|_m)\|\boldsymbol{u}\|_m. \end{aligned} \tag{3.8}$$

We shall next estimate the second term at the right-hand side of (3.6). In this regard, we notice that, by the Leibnitz rule, we have

$$-(\boldsymbol{v}\cdot\operatorname{grad}\boldsymbol{u},\boldsymbol{u})_m = -\sum_{0\le|\alpha|\le m}(\boldsymbol{v}\cdot\operatorname{grad}D^\alpha\boldsymbol{u},D^\alpha\boldsymbol{u})_0$$

$$-\sum_{\substack{1\le|\beta|\le|\alpha|\\0\le|\alpha|\le m}} C_{\alpha\beta}(D^\alpha\boldsymbol{v}\cdot\operatorname{grad}D^{\alpha-\beta}\boldsymbol{u},D^\alpha\boldsymbol{u})_0.$$

Since $\boldsymbol{v}\in\boldsymbol{X}_{m+2}$ we obtain after integration by parts

$$(\boldsymbol{v}\cdot\operatorname{grad}D^\alpha\boldsymbol{u},D^\alpha\boldsymbol{u})_0=0 \ \text{ for any multi-index } \alpha.$$

Moreover, again by the Sobolev embedding theorem,

$$-\sum_{\substack{1\le|\beta|\le|\alpha|\\0\le|\alpha|\le m}} C_{\alpha\beta}(D^\alpha\boldsymbol{v}\cdot\operatorname{grad}D^{\alpha-\beta}\boldsymbol{u},D^\alpha\boldsymbol{u})_0\le c\|\boldsymbol{v}\|_{m+2}\|\boldsymbol{u}\|_m^2,$$

and so we conclude

$$|(\boldsymbol{v}\cdot\operatorname{grad}\boldsymbol{u},\boldsymbol{u})_m|\le c\|\boldsymbol{v}\|_{m+2}\|\boldsymbol{u}\|_m^2. \tag{3.9}$$

From (3.6)-(3.9) it follows that

$$\frac{d}{dt}\|\boldsymbol{u}\|_m\le(c_1\|\boldsymbol{v}\|_{m+2}-\sigma)\|\boldsymbol{u}\|_m+c_2(\|\operatorname{grad}\tau\|_m+\|\boldsymbol{v}\|_{m+2}+\|\boldsymbol{F}\|_m). \tag{3.10}$$

It remains to estimate the term involving τ. To this end, by taking the divergence of both sides of (3.5), and recalling that $\boldsymbol{v}$ and $\boldsymbol{u}$ are solenoidal we derive that τ obeys the following Neumann problem at each $t_0\in I$

$$\Delta\tau=G \ \text{ in } \Omega$$

$$\frac{\partial\tau}{\partial n}=g \ \text{ at } \partial\Omega$$

where

$$G:=-\operatorname{div}(2\boldsymbol{u}\cdot\operatorname{grad}\boldsymbol{v}+\boldsymbol{u}\times\operatorname{curl}\boldsymbol{v}-\boldsymbol{N}(\boldsymbol{v})-\boldsymbol{F})$$

$$g:=-(\boldsymbol{u}\cdot\operatorname{grad}\boldsymbol{v}-\boldsymbol{u}\times\operatorname{curl}\boldsymbol{v}-\boldsymbol{N}(\boldsymbol{v})-\boldsymbol{F})\cdot\boldsymbol{n}.$$

In view of Lemma 2.1. and of a classical trace theorem (see, *e.g.* [11], Chapter II), we then find

$$\|\operatorname{grad}\tau\|_m\le c(\|\boldsymbol{v}\|_{m+1}\|\boldsymbol{u}\|_m+\|\boldsymbol{N}(\boldsymbol{v})\|_m+\|\boldsymbol{F}\|_m)$$

and so, by Lemma 3.1,

$$\|\operatorname{grad}\tau\|_m \leq c(\|\boldsymbol{v}\|_{m+1}\|\boldsymbol{u}\|_m + \|\boldsymbol{v}\|_{m+2} + \|\boldsymbol{F}\|_m). \tag{3.11}$$

Replacing this estimate into (3.10) we conclude

$$\frac{d}{dt}\|\boldsymbol{u}\|_m \leq (c_1\|\boldsymbol{v}\|_{m+2} + |\sigma|)\|\boldsymbol{u}\|_m + c_2(\|\boldsymbol{v}\|_{m+2} + \|\boldsymbol{F}\|_m). \tag{3.12}$$

Integrating this inequality over I with the help of Gronwall's lemma we find that

$$\|\boldsymbol{u}\|_m \leq D_1, \tag{3.13}$$

with $D_1 = D_1(\Omega, m, M, T, \nu, \alpha_1, \alpha_2, \|\boldsymbol{u}(t_0)\|_m, \|\boldsymbol{F}\|_{m,T})$. Furthermore, from (3.5), (3.11) and Lemma 3.1 we also find that

$$\begin{aligned}\left\|\frac{d\boldsymbol{u}}{dt}\right\|_{m-1} &\leq c(\|\boldsymbol{v}\|_m\|\boldsymbol{u}\|_{m-1} + \|\boldsymbol{v}\|_{m+1} + \|\boldsymbol{F}\|_{m-1} + \|\boldsymbol{u}\|_{m-1} \\ &+\|\boldsymbol{v}\cdot\operatorname{grad}\boldsymbol{u}\|_{m-1} + \|\boldsymbol{u}\cdot\operatorname{grad}\boldsymbol{v}\|_{m-1} + \|\boldsymbol{u}\times\operatorname{curl}\boldsymbol{v}\|_{m-1}).\end{aligned} \tag{3.14}$$

By the Sobolev embedding theorem, for all $k \geq 0$ we have

$$\|\boldsymbol{v}\|_{C^k} \leq c\|\boldsymbol{v}\|_{k+2}$$

which, in turn, once replaced into (3.14) furnishes

$$\begin{aligned}\left\|\frac{d\boldsymbol{u}}{dt}\right\|_{m-1} &\leq c(\|\boldsymbol{v}\|_m\|\boldsymbol{u}\|_{m-1} + \|\boldsymbol{v}\|_{m+1} + \|\boldsymbol{v}\|_{m+2}\|\boldsymbol{u}\|_m \\ &+\|\boldsymbol{F}\|_{m-1} + \|\boldsymbol{u}\|_{m-1}).\end{aligned}$$

Therefore, from the assumption made on $\boldsymbol{v}$ and (3.13) we conclude

$$\left\|\frac{d\boldsymbol{u}}{dt}\right\|_{m-1,T} \leq D_2 \tag{3.15}$$

with $D_2 = D_2(\Omega, m, M, T, \nu, \alpha_1, \alpha_2, \|\boldsymbol{u}(t_0)\|_m, \|\boldsymbol{F}\|_{m,T})$. Employing estimates (3.13) and (3.15), we can show the existence of a solution satisfying all requirements stated in the lemma. To this end, we shall use the Galerkin method with a special basis which we are going to introduce [22]. Let $\boldsymbol{g} \in \boldsymbol{X}_0$ and consider the problem

$$(\boldsymbol{\psi}, \boldsymbol{w})_m = (\boldsymbol{g}, \boldsymbol{w}), \quad \text{for all } \boldsymbol{v} \in \boldsymbol{X}_m. \tag{3.16}$$

By the Lax-Milgram theorem, there exists one and only one $\boldsymbol{\psi} \in X_m$ satisfying (3.15). In fact, by elliptic regularity, it can be shown [23] that $\boldsymbol{\psi} \in X_{2m}$. In view of the compact embedding

$$\boldsymbol{X}_m \hookrightarrow \boldsymbol{X}_0,$$

the linear map $\boldsymbol{g} \mapsto \boldsymbol{\psi}(\boldsymbol{g})$ is a compact, self-adjoint operator in $\boldsymbol{X}_0$ and it possesses an orthonormal complete family of eigenvectors $\boldsymbol{\psi}_k$ and corresponding eigenvalues λ_k:

$$\begin{aligned} &\boldsymbol{\psi}_k \in \boldsymbol{X}_{2m} \\ &(\boldsymbol{\psi}_k, \boldsymbol{w})_m = \lambda_k (\boldsymbol{\psi}_k, \boldsymbol{w}), \quad \text{for all } \boldsymbol{v} \in \boldsymbol{X}_m. \end{aligned} \tag{3.17}$$

We now look for an "approximating" solution to (3.1) of the form

$$\boldsymbol{u}_n(t) = \sum_{j=1}^{n} c_{jn}\boldsymbol{\psi}_k, \quad n \geq 0$$

$$\begin{aligned} \frac{d}{dt}(\boldsymbol{u}_n, \boldsymbol{\psi}_k) + \sigma((\boldsymbol{u}_n - \boldsymbol{v}), \boldsymbol{\psi}_k) + (\operatorname{curl} \boldsymbol{u}_n \times \boldsymbol{v}, \boldsymbol{\psi}_k) \\ = (\boldsymbol{N}(\boldsymbol{v}), \boldsymbol{\psi}_k) + (\boldsymbol{F}, \boldsymbol{\psi}_k), \quad k = 1, \ldots, n \end{aligned} \tag{3.18}$$

$$\boldsymbol{u}_n(t_0) = P_n \boldsymbol{u}(t_0),$$

where P_n is the orthogonal projection in $\boldsymbol{X}_0$ on the space spanned by the functions $\boldsymbol{\psi}_1, \ldots, \boldsymbol{\psi}_n$. For a given n, a solution to (3.18) exists in a time interval $[t_0, t_0 + T_n)$. It is easy to show that we can take $T_n = T$. In fact, multiplying (3.18_2) by $c_{kn}(t)$, adding in k, and using (3.4) we find

$$\begin{aligned} \tfrac{1}{2}\frac{d}{dt}\|\boldsymbol{u}_n\|^2 + \sigma\|\boldsymbol{u}_n\|^2 = \sigma(\boldsymbol{v}, \boldsymbol{u}_n) - (\boldsymbol{u}_n \cdot \operatorname{grad} \boldsymbol{v}, \boldsymbol{u}_n) \\ -(\boldsymbol{u}_n \times \operatorname{curl} \boldsymbol{v}, \boldsymbol{u}_n) + (\boldsymbol{N}(\boldsymbol{v}), \boldsymbol{u}_n) + (\boldsymbol{F}, \boldsymbol{u}_n). \end{aligned} \tag{3.19}$$

Using the embedding inequality

$$\|\operatorname{grad} \boldsymbol{v}\|_{L^\infty(\Omega)} \leq c\|\boldsymbol{v}\|_{m+2}, \quad m \geq 1,$$

the Schwarz and Cauchy inequalities along with Lemma 3.1, from (3.19) we deduce

$$\tfrac{1}{2}\frac{d}{dt}\|\boldsymbol{u}_n\|^2 \leq c(1 + \|\boldsymbol{v}\|_{m+2})\,\|\boldsymbol{u}_n\|^2 + \|\boldsymbol{v}\|^2_{m+2} + \|\boldsymbol{F}\|_m.$$

Thus, by Gronwall's lemma and the assumptions made on $\boldsymbol{v}$ and $\boldsymbol{F}$, we obtain

$$\|\boldsymbol{u}_m(t)\|^2 \leq \|\boldsymbol{u}_m(t_0)\|^2 e^{k_1(t-t_0)} + k_2\left(e^{k_1(t-t_0)} - 1\right) \tag{3.20}$$

with k_1 and k_2 depending only on M and $\|\boldsymbol{F}\|_{m,T}$. Taking into account that

$$\|\boldsymbol{u}_n(t_0)\|_m \leq \|\boldsymbol{u}(t_0)\|_m$$

from (3.20) we conclude, in particular, $T_n = T$, for all $n \geq 0$. Let us now show that $\{\boldsymbol{u}_n\}_{n\in\mathbb{N}}$ satisfies the estimates (3.13) and (3.15). To this end, we observe that, with the help of (3.4) and by the properties of the projection P (see Remark 2.1) equation (3.18$_2$) can be written as follows

$$\frac{d}{dt}(\boldsymbol{u}_n, \boldsymbol{\psi}_k) + \sigma((\boldsymbol{u}_n - \boldsymbol{v}), \boldsymbol{\psi}_k) \\ = -(P[\boldsymbol{v} \cdot \operatorname{grad} \boldsymbol{u}_n + \boldsymbol{u}_n \cdot \operatorname{grad} \boldsymbol{v} + \boldsymbol{u}_n \times \operatorname{curl} \boldsymbol{v} + \boldsymbol{N}(v) + \boldsymbol{F}], \boldsymbol{\psi}_k) \tag{3.21}$$

Recalling that $\boldsymbol{\psi} \in \boldsymbol{X}_{2m}$, by the assumptions made on $\boldsymbol{v}$ and $\boldsymbol{F}$, we easily recognize that a.e. in $[t_0, t_0 + T)$

$$(\boldsymbol{v} \cdot \operatorname{grad} \boldsymbol{u}_n + \boldsymbol{u}_n \cdot \operatorname{grad} \boldsymbol{v} + \boldsymbol{u}_n \times \operatorname{curl} \boldsymbol{v} - \boldsymbol{N}(v) - \boldsymbol{F}) \in \boldsymbol{H}^m(\Omega)$$

and, therefore,

$$P[\boldsymbol{v} \cdot \operatorname{grad} \boldsymbol{u}_n + \boldsymbol{u}_n \cdot \operatorname{grad} \boldsymbol{v} + \boldsymbol{u}_n \times \operatorname{curl} \boldsymbol{v} - \boldsymbol{N}(v) - \boldsymbol{F}] = \\ (\boldsymbol{v} \cdot \operatorname{grad} \boldsymbol{u}_n + \boldsymbol{u}_n \cdot \operatorname{grad} \boldsymbol{v} + \boldsymbol{u}_n \times \operatorname{curl} \boldsymbol{v} - \boldsymbol{N}(v) - \boldsymbol{F} + \operatorname{grad} \tau_n) \in \boldsymbol{X}_m \tag{3.22}$$

where τ_n solves the Neumann problem

$$\Delta \tau_n = G_n \quad \text{in } \Omega$$

$$\frac{\partial \tau_n}{\partial n} = g_n \quad \text{at } \partial\Omega$$

$$G_n := -\operatorname{div}(2\boldsymbol{u}_n \cdot \operatorname{grad} \boldsymbol{v} + \boldsymbol{u}_n \times \operatorname{curl} \boldsymbol{v} - \boldsymbol{N}(\boldsymbol{v}) - \boldsymbol{F})$$

$$g_n := -(\boldsymbol{u}_n \cdot \operatorname{grad} \boldsymbol{v} - \boldsymbol{u}_n \times \operatorname{curl} \boldsymbol{v} - \boldsymbol{N}(\boldsymbol{v}) - \boldsymbol{F}) \cdot \boldsymbol{n}.$$

Multiplying (3.21) by $\lambda_k c_{km}(t)$, summing over k and recalling (3.17), (3.22) we thus obtain

$$\tfrac{1}{2}\frac{d}{dt}\|\boldsymbol{u}_n\|_m^2 + \sigma\|\boldsymbol{u}_n\|_m^2 = \sigma(\boldsymbol{v}, \boldsymbol{u}_n)_m - (\boldsymbol{v} \cdot \operatorname{grad} \boldsymbol{u}_n, \boldsymbol{u}_n)_m - (\boldsymbol{u}_n \cdot \operatorname{grad} \boldsymbol{v}, \boldsymbol{u}_n)_m \\ -(\boldsymbol{u}_n \times \operatorname{curl} \boldsymbol{v}, \boldsymbol{u}_n)_m - (\operatorname{grad} \tau_n, \boldsymbol{u}_n)_m \\ +(\boldsymbol{N}(\boldsymbol{v}), \boldsymbol{u}_n)_m + (\boldsymbol{F}, \boldsymbol{u}_n)_m.$$

This equation coincides with (3.6) with $\boldsymbol{u}_n$ and τ_n in place of $\boldsymbol{u}$ and τ, respectively, and so, proceeding as before, we arrive at (3.13) (with $\boldsymbol{u}_n$ in place of $\boldsymbol{u}$). To show that $\boldsymbol{u}_n$ satisfies also (3.15) we begin to observe that multiplying (3.18$_2$) by dc_{kn}/dt and summing over k we find

$$\left\|\frac{\partial \boldsymbol{u}_n}{\partial t}\right\|^2 = -\sigma((\boldsymbol{u}_n - \boldsymbol{v}), \frac{\partial \boldsymbol{u}_n}{\partial t}) + (\operatorname{curl} \boldsymbol{u}_n \times \boldsymbol{v}, \frac{\partial \boldsymbol{u}_n}{\partial t}) \\ +(\boldsymbol{N}(\boldsymbol{v}), \frac{\partial \boldsymbol{u}_n}{\partial t}) + (\boldsymbol{F}, \frac{\partial \boldsymbol{u}_n}{\partial t}). \tag{3.23}$$

Using the Schwarz inequality at the right-hand side of this equation and recalling (3.13) (written for u_n) and the assumptions on v and F we obtain at once

$$\left\| \frac{\partial u_n}{\partial t} \right\|_{0,T} \leq C$$

with C independent of n. Thus, from this estimate and from (3.13) (written for u_n), using a standard procedure based on weak compactness we show that there exists a subsequence $\{u_{n'}\}_{n' \in \mathbb{N}}$ suitably converging to a field $u \in L^\infty(I; X_m) \cap W^{1,\infty}(I; X_0)$ which solves problem (3.1) a.e. We may then proceed as before to show the validity of estimate (3.15). The existence part of the lemma is then completed. Uniqueness is obtained easily by observing that, in view of the linearity of (3.5), it reduces to show that the problem

$$\left.\begin{aligned} &\frac{\partial u}{\partial t} + \sigma u + v \cdot \operatorname{grad} u = -u \cdot \operatorname{grad} v - u \times \operatorname{curl} v - \operatorname{grad} \tau \\ &\operatorname{div} u = 0 \end{aligned}\right\} \text{ in } \Omega_T$$

$$u \cdot n = 0 \quad \text{at } \partial\Omega \times (t_0, t_0 + T) \tag{3.24}$$

$$u(\cdot, t_0) = 0, \quad \text{in } \Omega,$$

admits the zero solution only. However, this is established at once, multiplying (3.24_1) by u, integrating by parts over Ω and using $(3.24_{2,3,4})$ and the regularity assumption on v along with Gronwall,s lemma.

Lemma 3.3. *Assume that*

$$v \in L^\infty(I, X_{m+2}), \quad m \geq 2, \quad \textit{with } \|v\|_{m+2,T} \leq CD,$$

$$F \in L^\infty(I, H_m) \textit{ with } \|F\|_{m,T} \leq \gamma D$$

$$u(t_0) \in X_m \textit{ with } \|u(t_0)\|_m \leq \beta D$$

where $C, \gamma, D > 0$ and $\beta < 1$. Then, if

$$T = \frac{1}{c_1 CD + |\sigma|} \ln \left(\frac{1 + \dfrac{c_2(c+\gamma)}{c_1 CD + |\sigma|}}{\beta + \dfrac{c_2(c+\gamma)}{c_1 CD + |\sigma|}} \right) \tag{3.25}$$

with c_1, c_2 suitable constants depending only on $\Omega, m, \alpha_1, \alpha_2$, the unique solution u, $\operatorname{grad} p$ to (3.1) determined in Lemma 3.2 *satisfies*

$$\|u\|_{m,T} \leq D.$$

Proof. We observe that the solution $\boldsymbol{u}$ obeys the differential inequality (3.12). Using Gronwall's lemma we find, in particular,

$$\|\boldsymbol{u}(t)\|_m \leq \left(\|\boldsymbol{u}(t_0)\|_m + \frac{k_2}{k_1}\right) e^{k_1 T} - \frac{k_2}{k_1},$$

with $k_1 = c_1 CD + |\sigma|$, $k_2 = c_2(C+\gamma)D$. Therefore, the result is a consequence of the choice (3.25).

Our next objective is to give existence and uniquenes for the boundary-value problem (3.3). Specifically, we have

Lemma 3.4. *Assume that* $m \geq 2$ *and*

$$\boldsymbol{v} \in \boldsymbol{X}_{m+2}, \ \textit{with} \ \|\boldsymbol{v}\|_{m+2} \leq CD,$$

$$\boldsymbol{F} \in \boldsymbol{H}^m(\Omega) \ \textit{with} \ \|\boldsymbol{F}\|_m \leq \beta D,$$

with $C > 0$. *Then, there exists a positive constant* $c = c(\Omega, m, \nu, \alpha_1, |\alpha_1 + \alpha_2|)$ *such that if*

$$D < c, \quad \beta < c,$$

problem (3.3) *admits a unique solution* $\boldsymbol{u}, \operatorname{grad} p$ *such that*

$$\boldsymbol{u} \in \boldsymbol{X}_m, \quad \operatorname{grad} p \in \boldsymbol{H}^m(\Omega).$$

Moreover,

$$\|\boldsymbol{u}\|_m \leq D.$$

Proof. We begin to derive an *a priori* estimate for the solutions to (3.3). To this end, we notice that, in view of the identity (3.4), equation (3.3_1) can be put in the following equivalent form

$$\nu \boldsymbol{u} + \alpha_1(\boldsymbol{v}\cdot\operatorname{grad}\boldsymbol{u} + \boldsymbol{u}\cdot\operatorname{grad}\boldsymbol{v} + \boldsymbol{u}\times\operatorname{curl}\boldsymbol{v}) - \boldsymbol{\omega}\times\boldsymbol{v} = -\operatorname{grad}\tau + \boldsymbol{N}(\boldsymbol{v}) + \boldsymbol{F}. \quad (3.26)$$

We start by deriving *a priori* estimates for problem (3.26)-$(3.2_{2,3})$. We apply the derivative operator D^k (k is a multi-index) to both sides of (3.26), take the scalar product in $L^2(\Omega)$ with $D^k\boldsymbol{u}$ and sum over k, with $0 \leq |k| \leq m$. We thus obtain

$$\begin{aligned} \nu\|\boldsymbol{u}\|_m &+ \alpha_1[(\boldsymbol{v}\cdot\operatorname{grad}\boldsymbol{u},\boldsymbol{u})_m + (\boldsymbol{u}\cdot\operatorname{grad}\boldsymbol{v},\boldsymbol{u})_m + (\boldsymbol{u}\times\operatorname{curl}\boldsymbol{v},\boldsymbol{u})_m] \\ &-(\boldsymbol{\omega}\times\boldsymbol{v},\boldsymbol{u})_m = -(\operatorname{grad}\tau,\boldsymbol{u})_m + (\boldsymbol{N}(\boldsymbol{v}),\boldsymbol{u})_m + (\boldsymbol{F},\boldsymbol{u})_m. \end{aligned} \quad (3.27)$$

From (3.27), the Schwarz inequality, (3.7), (3.9) and Lemma 3.1 we find

$$\|\boldsymbol{u}\|_m \leq c\left(\|\boldsymbol{v}\|_m\|\boldsymbol{u}\|_m + \|\boldsymbol{v}\|^2_{m+2} + \|\boldsymbol{F}\|_m + \|\operatorname{grad}\tau\|_m\right) \quad (3.28)$$

where c depends on $\alpha_1, |\alpha_1+\alpha_2|, \nu, \Omega$ and m. We need an estimate on the term involving τ. To this end, we proceed as in Lemma 3.2 and observe that from (3.26) it follows that τ obeys the following Neumann problem

$$\begin{aligned} \Delta\tau &= G \ \text{ in } \Omega \\ \frac{\partial\tau}{\partial n} &= g \ \text{ at } \partial\Omega \end{aligned} \tag{3.29$_1$}$$

where

$$\begin{aligned} G &:= -\operatorname{div}(2\boldsymbol{u}\cdot\operatorname{grad}\boldsymbol{v} + \boldsymbol{u}\times\operatorname{curl}\boldsymbol{v} - \boldsymbol{\omega}\times\boldsymbol{v} - \boldsymbol{N}(\boldsymbol{v}) - \boldsymbol{F}) \\ g &:= -(\boldsymbol{u}\cdot\operatorname{grad}\boldsymbol{v} - \boldsymbol{u}\times\operatorname{curl}\boldsymbol{v} - \boldsymbol{N}(\boldsymbol{v}) - \boldsymbol{F})\cdot\boldsymbol{n}. \end{aligned} \tag{3.29$_2$}$$

In view of Lemma 2.1. and of a classical trace theorem (see, *e.g.* [11], Chapter II), we then find

$$\|\operatorname{grad}\tau\|_m \le c(\|\boldsymbol{v}\|_{m+1}\|\boldsymbol{u}\|_m + \|\boldsymbol{v}\|_m^2 + \|\boldsymbol{N}(\boldsymbol{v})\|_m + \|\boldsymbol{F}\|_m)$$

and so, by Lemma 3.1, we obtain *a fortiori*

$$\|\operatorname{grad}\tau\|_m \le c(\|\boldsymbol{v}\|_{m+2}\|\boldsymbol{u}\|_m + \|\boldsymbol{v}\|_{m+2}^2 + \|\boldsymbol{F}\|_m).$$

Replacing this estimate into (3.28) we conclude

$$\|\boldsymbol{u}\|_m \le c\left(\|\boldsymbol{v}\|_m\|\boldsymbol{u}\|_m + \|\boldsymbol{v}\|_{m+2}^2 + \|\boldsymbol{F}\|_m\right).$$

By the assumptions made on $\boldsymbol{v}$ and $\boldsymbol{F}$, this latter inequality in turn implies

$$\|\boldsymbol{u}\|_m \le c\left(CD\|\boldsymbol{u}\|_m + C^2D^2 + \beta D\right)$$

from which is clear that if D and β satisfy the conditions stated in the lemma it follows that

$$\|\boldsymbol{u}\|_m \le D \tag{3.30}$$

which is the *a priori* estimate we wanted to obtain. Using the Galerkin method with the special basis (3.17) it is now simple to show the existence of a solution satisfying (3.29). We look for "approximating" solutions of the form

$$\begin{aligned} \boldsymbol{u}_n &= \sum_{j=1}^{n} c_{jn}\boldsymbol{\psi}_k, \quad n \ge 0 \\ \nu(\boldsymbol{u}_n, \boldsymbol{\psi}_k) &+ (P[\alpha_1\operatorname{curl}\boldsymbol{u}_n\times\boldsymbol{v} - \boldsymbol{\omega}\times\boldsymbol{v}], \boldsymbol{\psi}_k) \\ &= (P[\boldsymbol{N}(\boldsymbol{v}) + \boldsymbol{F}], \boldsymbol{\psi}_k), \quad k = 1, \dots, n \end{aligned} \tag{3.31}$$

where P is the orthogonal projection of $L^2(\Omega)$ onto $\boldsymbol{X}_0$. Existence to the algebraic system (3.30) in the unknowns c_{kn} will follow from Brower fixed point theorem as soon as we show that, denoting by $\mathcal{X}_n$ the space spanned by $\boldsymbol{\psi}_1, \ldots, \boldsymbol{\psi}_n$, the map

$$\mathcal{L} : \boldsymbol{c} \equiv (c_{1n}, \ldots, c_{nn}) \in \mathcal{X}_n \mapsto \mathcal{L}(\boldsymbol{c}) \in \mathcal{X}_n$$

with

$$\begin{aligned} \mathcal{L}_k(\boldsymbol{c}) = &- (P[\alpha_1 \operatorname{curl} \boldsymbol{u}_n \times \boldsymbol{v} - \boldsymbol{\omega} \times \boldsymbol{v}], \boldsymbol{\psi}_k) \\ &+ (P[\boldsymbol{N}(\boldsymbol{v}) + \boldsymbol{F}], \boldsymbol{\psi}_k), \quad k = 1, \ldots, n \end{aligned} \tag{3.32}$$

admits the following estimate

$$\mathcal{L}(\boldsymbol{c}) \cdot \boldsymbol{c} \leq \rho \|\boldsymbol{u}_n\| + \delta \|\boldsymbol{u}_n\|^2 \tag{3.33}$$

for some $\rho > 0$ and $\delta < \nu$ (see [12], Lemma VIII.3.1). From (3.32) and (3.4) we find

$$\mathcal{L}(\boldsymbol{c}) \cdot \boldsymbol{c} = -\alpha_1(\boldsymbol{u}_n \cdot \operatorname{grad} \boldsymbol{v} + \boldsymbol{u}_n \times \operatorname{curl} \boldsymbol{v}, \boldsymbol{u}_n) + (\boldsymbol{N}(\boldsymbol{v}) + \boldsymbol{F}, \boldsymbol{u}_n)$$

and so, using the Schwarz inequality in this latter equation, by the assumption made on D, condition (3.33) follows. To show (3.30), we multiply (3.31) by $\lambda_k c_{kn}$, sum over k and use (3.17) and (3.4) to find

$$\begin{aligned} \nu \|\boldsymbol{u}_n\|_m + \alpha_1 [(\boldsymbol{v} \cdot \operatorname{grad} \boldsymbol{u}_n, \boldsymbol{u}_n)_m + (\boldsymbol{u}_n \cdot \operatorname{grad} \boldsymbol{v}, \boldsymbol{u}_n)_m + (\boldsymbol{u}_n \times \operatorname{curl} \boldsymbol{v}, \boldsymbol{u}_n)_m] \\ - (\boldsymbol{\omega} \times \boldsymbol{v}, \boldsymbol{u}_n)_m = -(\operatorname{grad} \tau_n, \boldsymbol{u}_n)_m + (\boldsymbol{N}(\boldsymbol{v}), \boldsymbol{u}_n)_m + (\boldsymbol{F}, \boldsymbol{u}_n)_m \end{aligned} \tag{3.34}$$

where τ_n solves $(3.29_{1,2})$ with $\boldsymbol{u}_n$ in place of $\boldsymbol{u}$. Now, (3.34) coincides with (3.28) with $\boldsymbol{u}_n$ in place of $\boldsymbol{u}$ and so, proceeding as before, we arrive at (3.30). Using (3.30) for $\boldsymbol{u}_n$, the compactness property of the space $\boldsymbol{H}^m(\Omega)$ and the assumption made on $\boldsymbol{v}$, it is routine to show that there exists a subsequence $\{\boldsymbol{u}_{n'}\}$ suitably converging to a field $\boldsymbol{u}$ which satisfies all the properties stated in the lemma, thus completing the existence part. For uniqueness, it is enough to show that the problem

$$\left.\begin{aligned} &\nu \boldsymbol{u} + \alpha_1(\boldsymbol{v} \cdot \operatorname{grad} \boldsymbol{u} + \boldsymbol{u} \cdot \operatorname{grad} \boldsymbol{v} + \boldsymbol{u} \times \operatorname{curl} \boldsymbol{v}) = -\operatorname{grad} \tau \\ &\operatorname{div} \boldsymbol{u} = 0 \end{aligned}\right\} \text{ in } \Omega \tag{3.35}$$

$$\boldsymbol{u} \cdot \boldsymbol{n} = 0 \quad \text{at } \Omega$$

admits only the zero solution. Multiplying (3.35_1) by $\boldsymbol{u}$, integrating over Ω and using $(3.35_{2,3})$, we find

$$\nu \|\boldsymbol{u}\|^2 = -\alpha_1(\boldsymbol{u} \cdot \operatorname{grad} \boldsymbol{v} + \boldsymbol{u} \times \operatorname{curl} \boldsymbol{v}, \boldsymbol{u})$$

and so, by the smallness assumption on D we easily conclude $\boldsymbol{u} \equiv 0$. The lemma is therefore completely proved.

4. Existence and Uniqueness for the $\mathcal{IBVP}$.

In this section we shall show that the $\mathcal{IBVP}$ (1.7)-(1.11),(1.13) admits a unique classical solution in $[t_0, t_0+T)$ for all $t_0 \geq 0$ and sufficiently small T depending on the size of the initial data. For this result to hold, we don't need any assumption on $|\alpha_1+\alpha_2|$, while α_1 is assumed positive or even negative, provided it ranges in a certain interval of negative numbers. We also show that it is possible to take $T=\infty$, thus obtaining a *global* solution, if the size of the initial data is suitably restricted. In such a case, the hypothesis $\alpha > 0$ is crucial.

These results will be obtained by means of a fixed point procedure that we are about to describe. Applying the Helmholtz decomposition (see Lemma 2.2) to the field $\boldsymbol{v}-\alpha_1\Delta\boldsymbol{v}$, and recalling (1.7), (1.13) we have

$$\left.\begin{aligned}&\boldsymbol{v}-\alpha_1\Delta\boldsymbol{v}=\boldsymbol{u}+\operatorname{grad}\pi\\&\operatorname{div}\boldsymbol{v}=0\end{aligned}\right\}\quad\text{in } \Omega_T$$
$$\boldsymbol{v}=0\quad\text{at } \partial\Omega\times I$$

where, taking also into account (1.8), the field $\boldsymbol{u}$ satisfies

$$\left.\begin{aligned}&\frac{\partial\boldsymbol{u}}{\partial t}+\sigma(\boldsymbol{u}-\boldsymbol{v})+\operatorname{curl}\boldsymbol{u}\times\boldsymbol{v}=-\operatorname{grad}p+\boldsymbol{N}(\boldsymbol{v})+\boldsymbol{F}\\&\operatorname{div}\boldsymbol{u}=0\end{aligned}\right\}\quad\text{in } \Omega_T$$
$$\boldsymbol{u}\cdot\boldsymbol{n}=0\quad\text{at } \partial\Omega\times I \tag{4.1}$$
$$\boldsymbol{u}(\cdot,t_0)=\boldsymbol{u}(t_0),\quad\text{in } \Omega,$$

where $\sigma=\nu/\alpha_1$, $p=P+\dfrac{\partial\pi}{\partial t}$, and $\boldsymbol{N}(\boldsymbol{v})$ is defined in (3.2). Take the Banach space

$$Y=C(I;\boldsymbol{X}_{m-1}),\qquad m\geq 1$$

and, for $D>0$, define

$$G=\Big\{\varphi\in Y\colon\ \varphi\in L^\infty(I;\boldsymbol{H}^m(\Omega)),\ \|\varphi\|_{m,T}\leq D,\ \varphi(x,t_0)=\boldsymbol{u}(x,t_0)\in\boldsymbol{X}_m\Big\}.$$

Consider now the map Φ defined in G, in the following way:

$$\Phi\colon\ \varphi\mapsto\boldsymbol{u}$$

is the composition of the operator $\varphi\mapsto\boldsymbol{v}$ defined by

$$\left.\begin{aligned}&\boldsymbol{v}-\alpha_1\,\Delta\boldsymbol{v}=\varphi+\operatorname{grad}\pi\\&\operatorname{div}\boldsymbol{v}=0\end{aligned}\right\}\quad\text{in } \Omega_T \tag{4.2}$$
$$\boldsymbol{v}=0\quad\text{at } \partial\Omega\times I$$

(where the time t is a parameter), with the operator $\boldsymbol{v} \mapsto \boldsymbol{u}$, defined by (4.1). It is clear that to prove the existence of a solution to our original problem (1.7)-(1.11),(1.13) is equivalent to show that the map

$$\Phi: \ G \subset Y \ \rightarrow Y$$

admits a fixed point. The existence of such a fixed point will be proved by means of the following theorem [10].

Lemma 4.1 (Schauder Fixed Point). *A compact mapping Φ of a closed bounded convex set G in a Banach space Y into itself has a fixed point.*

First we prove the following

Lemma 4.2. *Let*

$$1/\alpha > -\lambda_1$$

with λ_1 defined before Lemma 2.3. *Then, for all t_0 and D the map Φ transforms the closed bounded convex set G into a relatively compact subset of Y. Moreover, Φ is continuous in the topology of Y.*

Proof. The closedness of set G is obvious. In fact every sequence φ_n $(n = 1, ...)$ in G, converging to φ in Y has a subsequence φ_{n_k} which converges weakly to a certain ψ in $L^\infty(I; \boldsymbol{H}^m(\Omega))$ such that $\|\psi\|_{m,T} \leq \liminf \|\varphi_{n_k}\|_{m,T}$. Thus $\|\varphi\|_{m,T} = \|\psi\|_{m,T} \leq D$. To prove the compactness property of Φ, let us denote by $\boldsymbol{v}_n$, a sequence of solutions to problem (4.2), corresponding to the data $\varphi_n \in G$, such that $\boldsymbol{v}_n$ is uniformly bounded on I, for the $\boldsymbol{H}^{m+2}(\Omega)$-norm (Lemma 2.3). Let $\boldsymbol{u}_n \in G$ be the associated sequence of solutions to problem (4.1). Since

$$\boldsymbol{u}_n \quad \text{is bounded in } L^\infty(I; \boldsymbol{V}_m)\,,$$

$$\frac{d\boldsymbol{u}_n}{dt} \quad \text{is bounded in } L^\infty(I; \boldsymbol{X}_{m-1})\,,$$

we have in particular that $\boldsymbol{u}_n$ is bounded in $W^{1,2}(I; \boldsymbol{X}_{m-1})$ and, by classical compactness arguments we conclude that $\boldsymbol{u}_n \rightarrow \boldsymbol{u}$ in Y. Concerning the continuity of Φ we still denote by $\boldsymbol{u}, \boldsymbol{u}_n \in Y$ $(n = 1, ...)$ the corresponding images of $\varphi, \varphi_n \in G$ $(n = 1, ...)$ by the map Φ. Let us subtract the equations $(1.9)_1$ written for $\boldsymbol{u}$ and $\boldsymbol{u}_n$, multiply by $\boldsymbol{u}_n - \boldsymbol{u}$ and calculate the $\boldsymbol{H}^{m-1}(\Omega)$-inner product. Setting $y(t) = \|\boldsymbol{u}_n - \boldsymbol{u}\|_{m-1}$ we get, after some easy calculations

$$y'(t) \leq \lambda\, y(t) + C\Big(1 + \|\boldsymbol{u}\|_{m-1,T} + \|\boldsymbol{u}\|_{m,T}\Big)\, \|\varphi_n - \varphi\|_Y\,,$$

with $\lambda \geq 0$. Then noticing that $\|\boldsymbol{u}\|_m$ is bounded (Lemma 3.2) we conclude that $\|\boldsymbol{u}_n - \boldsymbol{u}\|_Y \leq C_1 \|\varphi_n - \varphi\|_Y$ and thus the continuity of Φ is proved.

We are now ready to prove

Theorem 4.1 (Local Existence). *Let Ω be of class C^{m+2}, $m \geq 2$ and let*

$$1/\alpha > -\lambda_1$$

with λ_1 defined before Lemma 2.3. *Then, for any $t_0 \geq 0$ and $\boldsymbol{v}(x, t_0) \in \boldsymbol{X}_{m+2}$, $m \geq 2$, problem* (1.7)-(1.11), (1.13) *has a unique solution in $[t_0, t_0 + T]$ with T given in* (3.25) *such that*

$$\begin{aligned} &\boldsymbol{v} \in C(I; \boldsymbol{X}_{m+1}) \cap L^\infty(I; \boldsymbol{H}^{m+2}(\Omega)), \quad \operatorname{grad} P \in L^\infty(I; \boldsymbol{H}^{m+1}(\Omega)) \\ &\frac{d\boldsymbol{v}}{dt} \in L^\infty(I; \boldsymbol{H}^{m+1}(\Omega)) \,. \end{aligned} \tag{4.3}$$

In particular, if $m = 5$ and if, in addition, $\boldsymbol{F} \in W^{1,\infty}(I, \boldsymbol{H}^3(\Omega))$, then

$$\boldsymbol{v} \in C^1(I; C^3(\Omega)), \quad P \in C(I; C^1(\Omega))$$

and the solution is classical.

Proof. By what we said, to show the first part of the theorem it is enough to show taht Φ has a fixed point in G. To this end, we apply Lemma 4.1. In view of Lemma 4.2, for existence and the proof of (4.3) we only need to prove that Φ maps G into itself. However, this is an immediate consequence of Lemma 3.3. Let us now prove the regularity of $\boldsymbol{v}$. This will follow as soon as we show that the solution $\boldsymbol{u}$ to (4.1) is in $W^{2,\infty}(I; \boldsymbol{X}_3)$, for this would yield, by Lemma 2.3, that

$$\boldsymbol{v} \in W^{2,\infty}(I; \boldsymbol{X}_5) \hookrightarrow C^1(I; C^3(\Omega)).$$

We therefore need to prove that

$$\frac{d^2\boldsymbol{u}}{dt^2} \in L^\infty(I; \boldsymbol{X}_3). \tag{4.4}$$

To this end, formally differentiating (4.1) with respect to time we obtain (with subscript t meaning partial differentiation)

$$\boldsymbol{u}_{tt} + \sigma(\boldsymbol{u} - \boldsymbol{v})_t + (\operatorname{curl} \boldsymbol{u})_t \times \boldsymbol{v} + \operatorname{curl} \boldsymbol{u} \times \boldsymbol{v}_t = -\operatorname{grad} p_t + (\boldsymbol{N}(\boldsymbol{v}))_t + \boldsymbol{F}_t. \tag{4.5}$$

From the first part of the theorem and the definition of $\boldsymbol{N}(\boldsymbol{v})$) it easily follows that

$$\sigma(\boldsymbol{u} - \boldsymbol{v})_t + (\operatorname{curl} \boldsymbol{u})_t \times \boldsymbol{v} + \operatorname{curl} \boldsymbol{u} \times \boldsymbol{v}_t - (\boldsymbol{N}(\boldsymbol{v}))_t + \boldsymbol{F}_t \in L^\infty(I; \boldsymbol{H}^3(\Omega)). \tag{4.6}$$

In order to estimate the term $(\operatorname{grad} p)_t$ we consider the Neumann problem

$$\begin{aligned} &\Delta p = -\operatorname{div}[(\operatorname{curl} \boldsymbol{u}) \times \boldsymbol{v} - \boldsymbol{N}(\boldsymbol{v}) + \boldsymbol{F}] := \operatorname{div} \boldsymbol{G} \;\text{ in } \Omega_T \\ &\frac{\partial p}{\partial n} = \boldsymbol{G} \cdot \boldsymbol{n} := H \;\text{ at } \partial\Omega \times I. \end{aligned}$$

Now, since

$$G_t \in L^\infty(I; \boldsymbol{H}^3(\Omega)), \quad H_t \in L^\infty(I; H^3(\Omega)),$$

we can find a solution to

$$\Delta\varphi = \operatorname{div} \boldsymbol{G}_t \quad \text{in } \Omega_T$$

$$\frac{\partial \varphi}{\partial n} = H_t \quad \text{at } \partial\Omega \times I$$

which satisfies $\operatorname{grad}\varphi \in L^\infty(I; \boldsymbol{H}^3(\Omega))$. A starightforward distribution-theoretical argument shows that $\Delta\varphi$ is the distributional time derivative of Δp, and that $(\operatorname{grad} p)_t \in L^\infty(I; \boldsymbol{H}^3(\Omega))$. This latter property together with (4.5), (4.6) implies (4.4) and the proof of existence is completed. To show uniqueness, we denote by $\boldsymbol{v}_1, P_1$ and $\boldsymbol{v}_2, P_2$ two solutions corresponding to the same data and set

$$\boldsymbol{w} = \boldsymbol{v}_1 - \boldsymbol{v}_2, \quad \boldsymbol{U} = P(\boldsymbol{w} - \alpha_1 \Delta \boldsymbol{w}),$$

where P is the projection operator of $\boldsymbol{L}^2(\Omega)$ into $\boldsymbol{X}_0$. From (4.1) and (3.4) it follows that

$$\frac{\partial \boldsymbol{U}}{\partial t} + \sigma(\boldsymbol{U} - \boldsymbol{w}) + \boldsymbol{v}_1 \cdot \operatorname{grad} \boldsymbol{U} + \boldsymbol{U} \cdot \operatorname{grad} \boldsymbol{v}_1$$

$$+\boldsymbol{U} \times \operatorname{curl} \boldsymbol{v}_1 + \operatorname{curl} \boldsymbol{u}_1 \times \boldsymbol{w} = -\operatorname{grad} P' + \boldsymbol{N}(\boldsymbol{v}_1) - \boldsymbol{N}(\boldsymbol{v}_2).$$

Multiplying both sides of this equation by $\boldsymbol{U}$, integrating by parts over Ω and recalling that the normal component of $\boldsymbol{U}$ vanishes at the boundary, we find

$$\frac{1}{2}\frac{d}{dt}\|\boldsymbol{U}\|^2 + \sigma\|\boldsymbol{U}\|^2 = \sigma(\boldsymbol{w}, \boldsymbol{U}) - (\boldsymbol{U} \cdot \operatorname{grad} \boldsymbol{v}_1, \boldsymbol{U}) - (\boldsymbol{U} \times \operatorname{curl} \boldsymbol{v}_1, \boldsymbol{U})$$

$$-(\operatorname{curl} \boldsymbol{u}_1 \times \boldsymbol{w}, \boldsymbol{U}) + (\boldsymbol{N}(\boldsymbol{v}_1) - \boldsymbol{N}(\boldsymbol{v}_2), \boldsymbol{U}).$$

Employing the regularity properties of $\boldsymbol{v}_1$, the Schwarz inequality and Lemmas 2.3 and 3.1, from the preceding relation we obtain

$$\frac{d}{dt}\|\boldsymbol{U}\|^2 \le c\|\boldsymbol{U}\|^2,$$

with $c = c(\boldsymbol{v}_1, \boldsymbol{v}_2)$. Recalling that $\boldsymbol{U}(x, 0) = 0$ for all $x \in \Omega$, by Gronwall's lemma we conclude $\boldsymbol{U} \equiv 0$ which, in turn, by Lemma 2.3 implies $\boldsymbol{w} \equiv 0$ and uniqueness follows.

Our next objective is to show that if α_1 is *strictly* positive and the size of the data is sufficiently small, the solution determined in Theorem 4.1 exists for all $T > 0$. This will be obtained by means of a "boot-strap" argument based on the local existence result just shown and on global estimates which we are going to derive. We wish to emphasize that we are able to show these estimates only under the assumption that Ω is simply connected. Therefore, the general case is left as an open problem.

Lemma 4.3. *Let Ω be simply connected of class C^{m+2}, and let $\boldsymbol{v}$ be a solution to (1.7)-(1.11), (1.13) such that*

$$\boldsymbol{v} \in L^\infty(0,T;\boldsymbol{X}_{m+2}) \cap W^{1,\infty}(0,T;\boldsymbol{X}_{m+1}).$$

Assume, further,

$$\boldsymbol{v}(0) \in \boldsymbol{X}_{m+2}, \quad \boldsymbol{F} \in L^\infty(0,T;\boldsymbol{H}^m(\Omega)).$$

Then, if $m \geq 1$, there exists $\delta = \delta(\Omega, m, \nu, \alpha_1, \alpha_2) > 0$ such that if

$$\|\boldsymbol{v}_0\|_{m+2}^2 + \int_0^T \|\boldsymbol{F}\|_m^2 \leq \delta \tag{4.7}$$

we have

$$\|\boldsymbol{v}\|_{m+2,T}^2 + \int_0^T \|\boldsymbol{v}(s)\|_{m+2}^2 \leq \delta_1 \left(\|\boldsymbol{v}(0)\|_{m+2}^2 + \int_0^T \|\boldsymbol{F}\|_{m,T}^2 \right). \tag{4.8$_1$}$$

Furthermore, if $m \geq 2$, then

$$\left\| \frac{d\boldsymbol{v}}{dt} \right\|_{m+1,T} \leq \delta_2. \tag{4.8$_2$}$$

In these relations the quantities δ_1 and δ_2 depend only on $\Omega, m, \nu, \alpha_1, \alpha_2$ and δ.

Proof. Let us multiply (1.8) by $\boldsymbol{v}$ and integrate by parts over Ω. Taking into account (1.7) and (1.13) we deduce

$$\tfrac{1}{2}\frac{d}{dt}\left(\|\boldsymbol{v}\|^2 + \alpha_1\|\operatorname{grad}\boldsymbol{v}\|^2\right) = -\nu\|\operatorname{grad}\boldsymbol{v}\|^2 + (\boldsymbol{N}(\boldsymbol{v}),\boldsymbol{v}) + (\boldsymbol{F},\boldsymbol{v}). \tag{4.10}$$

We now apply the curl operator at both sides of (4.1$_1$), and use the identity

$$\operatorname{curl}(\boldsymbol{A}\times\boldsymbol{B}) = \boldsymbol{B}\cdot\operatorname{grad}\boldsymbol{A} - \boldsymbol{A}\cdot\operatorname{grad}\boldsymbol{B} + (\operatorname{div}\boldsymbol{B})\,\boldsymbol{A} - (\operatorname{div}\boldsymbol{A})\,\boldsymbol{B}$$

to deduce

$$\frac{\partial \boldsymbol{U}}{\partial t} + \sigma(\boldsymbol{U} - \operatorname{curl}\boldsymbol{v}) = \boldsymbol{U}\cdot\operatorname{grad}\boldsymbol{v} - \boldsymbol{v}\cdot\operatorname{grad}\boldsymbol{U} + \operatorname{curl}\boldsymbol{N}(\boldsymbol{v}) + \operatorname{curl}\boldsymbol{F}, \tag{4.11}$$

where $\boldsymbol{U} := \operatorname{curl}\boldsymbol{u}$. Multiplying this equation by $\boldsymbol{U}$ and integrating by parts, it follows that

$$\begin{aligned}\tfrac{1}{2}\frac{d}{dt}\|\boldsymbol{U}\|^2 + \sigma\|\boldsymbol{U}\|^2 &= \sigma(\operatorname{curl}\boldsymbol{v},\boldsymbol{U}) + (\boldsymbol{U}\cdot\operatorname{grad}\boldsymbol{v},\boldsymbol{U}) \\ &\quad + (\operatorname{curl}\boldsymbol{N}(\boldsymbol{v}),\boldsymbol{U}) + (\operatorname{curl}\boldsymbol{F},\boldsymbol{U}).\end{aligned} \tag{4.12}$$

By the Cauchy inequality we have

$$|(\operatorname{curl} \boldsymbol{v}, \boldsymbol{U})| \leq \tfrac{1}{2}\|\boldsymbol{U}\|^2 + \tfrac{1}{2}\|\operatorname{grad} \boldsymbol{v}\|^2$$

and so, using this relation into the (4.11), multiplying (4.9) by $\lambda > 0$ and adding the two displayed relations, we obtain

$$\begin{aligned}\tfrac{1}{2}\frac{d}{dt}(\|\boldsymbol{v}\|^2 + \alpha_1\|\operatorname{grad}\boldsymbol{v}\|^2 + \lambda\|\boldsymbol{U}\|^2) &\leq -\tfrac{1}{2}(\sigma\|\boldsymbol{U}\|^2 + \tau\|\operatorname{grad}\boldsymbol{v}\|^2)\\ &+\lambda[(\boldsymbol{U}\cdot\operatorname{grad}\boldsymbol{v},\boldsymbol{U}) + (\operatorname{curl}\boldsymbol{N}(\boldsymbol{v}),\boldsymbol{U}) + (\operatorname{curl}\boldsymbol{F},\boldsymbol{U})]\\ &+(\boldsymbol{N}(\boldsymbol{v}),\boldsymbol{v}) + (\boldsymbol{F},\boldsymbol{v})\end{aligned} \tag{4.13}$$

with $\tau = 2\nu - \sigma\lambda$. We choose λ such that $\tau > 0$. By the Sobolev embedding theorem and by Lemmas 2.4 and 3.1 it follows that

$$\begin{aligned}|(\boldsymbol{U}\cdot\operatorname{grad}\boldsymbol{v},\boldsymbol{U})| &\leq c\|\boldsymbol{U}\|^2\|\boldsymbol{v}\|_3 \leq c\|\boldsymbol{U}\|^3\\ |(\operatorname{curl}\boldsymbol{N}(\boldsymbol{v}),\boldsymbol{U})| &\leq c\|\boldsymbol{v}\|_3^2\|\boldsymbol{U}\| \leq c\|\boldsymbol{U}\|^3\\ |(\operatorname{curl}\boldsymbol{F},\boldsymbol{U})| + |(\boldsymbol{F},\boldsymbol{v})| &\leq \|\boldsymbol{F}\|_1(\|\boldsymbol{v}\| + \|\boldsymbol{U}\|)\\ |(\boldsymbol{N}(\boldsymbol{v}),\boldsymbol{v})| &\leq c\|\boldsymbol{v}\|_3^2\|\boldsymbol{v}\| \leq c\|\boldsymbol{U}\|^3.\end{aligned}$$

Replacing these inequalities into (4.12), after a simple manipulation we deduce

$$\frac{d}{dt}[\![\boldsymbol{v}]\!]^2 \leq -\chi[\![\boldsymbol{v}]\!]^2 + c[\![\boldsymbol{v}]\!]^3 + \|\boldsymbol{F}\|_1^2 \tag{4.14}$$

where

$$[\![\boldsymbol{v}]\!]^2 := \|\boldsymbol{v}\| + \alpha_1\|\operatorname{grad}\boldsymbol{v}\| + \lambda\|\boldsymbol{U}\|$$

and $\chi, c > 0$. Notice that, in view of Lemma 2.4, $[\![\boldsymbol{v}]\!]$ is equivalent to $\|\boldsymbol{v}\|_3$. Thus, we use Lemma 2.5 into (4.13) to obtain $\|\boldsymbol{v}\|_{3,T} \leq \sigma_1$, with σ_1 controlled by δ. Putting this information back into (4.13), for sufficiently small σ_1 (*i.e.* δ) we find

$$\frac{d}{dt}[\![\boldsymbol{v}]\!]^2 \leq -\chi'[\![\boldsymbol{v}]\!]^2 + \|\boldsymbol{F}\|_1^2,$$

for some $\chi' > 0$, which, in turn, upon integration, proves (4.8_1) for $m = 1$. Let us show the general case by induction. Thus, assuming that (4.8_1) holds for m, let us show that it holds for $m+1$ too. We apply the operator D^k to both sides of (4.10), multiply the resulting equation by $D^k\boldsymbol{U}$ and integrate over Ω. We thus find

$$\begin{aligned}\frac{d}{dt}\|\boldsymbol{U}\|_m^2 + \sigma\|\boldsymbol{U}\|_m^2 = &\,\sigma(\operatorname{curl}\boldsymbol{v},\boldsymbol{U})_m + (\boldsymbol{U}\cdot\operatorname{grad}\boldsymbol{v},\boldsymbol{U})_m - (\boldsymbol{v}\cdot\operatorname{grad}\boldsymbol{U},\boldsymbol{U})_m\\ &+(\operatorname{curl}\boldsymbol{N}(\boldsymbol{v}),\boldsymbol{U})_m + (\operatorname{curl}\boldsymbol{F},\boldsymbol{U})_m.\end{aligned} \tag{4.15}$$

Using the Cauchy and Schwarz inequalities, Lemmas 2.4 and 3.1, and inequality (3.9) (with $\boldsymbol{U}$ in place of $\boldsymbol{u}$), we obtain

$$|(\operatorname{curl}\boldsymbol{v},\boldsymbol{U})_m| \le \tfrac{1}{2}\left(\|\boldsymbol{v}\|_{m+1}^2 + \|\boldsymbol{U}\|_m^2\right)$$

$$|(\boldsymbol{U}\cdot\operatorname{grad}\boldsymbol{v},\boldsymbol{U})_m| \le c\|\boldsymbol{v}\|_{m+2}\|\boldsymbol{U}\|_m^2 \le c\|\boldsymbol{U}\|_m^3$$

$$|(\operatorname{curl}\boldsymbol{N}(\boldsymbol{v}),\boldsymbol{U})_m| \le c\|\boldsymbol{v}\|_{m+2}^2\|\boldsymbol{U}\|_m \le c\|\boldsymbol{U}\|_m^3$$

$$|(\operatorname{curl}\boldsymbol{F},\boldsymbol{U})_m| \le c\|\boldsymbol{F}\|_{m+1}^2 + \frac{\sigma}{4}\|\boldsymbol{U}\|_m^2.$$

Moreover, by the Sobolev embedding theorem, we deduce

$$|(\boldsymbol{v}\cdot\operatorname{grad}\boldsymbol{U},\boldsymbol{U})_m| \le c\|\boldsymbol{v}\|_{m+2}\|\boldsymbol{U}\|_m^2 \le c\|\boldsymbol{U}\|_m^3.$$

Replacing these estimates into (4.14) we conclude

$$\frac{d}{dt}\|\boldsymbol{U}\|_m^2 + \chi\|\boldsymbol{U}\|_m^2 \le c\|\boldsymbol{U}\|_m^3 + \mathcal{F} \tag{4.16}$$

where

$$\mathcal{F} = \|\boldsymbol{v}(t)\|_{m+1}^2 + \|\boldsymbol{F}(t)\|_{m+1}^2$$

and $\chi > 0$. The inequality (4.8_1) then becomes a consequence of (4.15), Lemma 2.5 and the induction hypothesis. The proof of (4.8_2) is at once obtained from (4.10) with the help of (4.8_1) and Lemmas 2.4 and 3.1. The lemma is proved.

Using the global *a priori* estimates of this lemma, we can extend the local existence of Theorem 4.1 to deduce the following main result.

Theorem 4.2 (Global Existence). *Let Ω be simply connected and of class C^{m+2}, $m \ge 2$. Moreover, let*

$$\boldsymbol{v}_0 \in \boldsymbol{X}_{m+2}, \quad \boldsymbol{F} \in L^\infty(0,T;\boldsymbol{H}^m(\Omega)), \quad m \ge 2, \ T > 0.$$

Then, there exists $\delta_0 = \delta_0(\Omega, m, \nu, \alpha_1, \alpha_2) > 0$ such that, if

$$\mathcal{D} := \|\boldsymbol{v}_0\|_{m+2} + \int_0^T \|\boldsymbol{F}(t)\|_m^2 \le \delta_0,$$

problem (1.7)-(1.11), (1.13) has a unique solution for all $t \in [0,T)$ such that

$$\begin{aligned} &\boldsymbol{v} \in C(0,T;\boldsymbol{X}_{m+1}) \cap L^\infty(0,T;\boldsymbol{H}^{m+2}(\Omega)), \quad \operatorname{grad} P \in L^\infty(0,T;\boldsymbol{H}^{m+1}(\Omega)) \\ &\frac{d\boldsymbol{v}}{dt} \in L^\infty(0,T;\boldsymbol{H}^{m+1}(\Omega)). \end{aligned} \tag{4.17}$$

Moreover, for $m \ge 3$

$$\frac{d^2\boldsymbol{v}}{dt^2} \in L^\infty(0,T;\boldsymbol{H}^m(\Omega))$$

Thus in particular, for $m = 5$, $\boldsymbol{v}$ is a classical solution, i.e.

$$\boldsymbol{v} \in C^1(0,T;\boldsymbol{C}^3(\Omega)), \quad P \in C(0,T;C^1(\Omega)).$$

Proof. We need to show only the first part (global existence), the second one being proved exactly as in Theorem 4.1. We choose $\mathcal{D} \le \delta = \varepsilon$. By the Theorem 4.1 we have the existence on $[0, T_1]$. with T_1 given by (3.25). Notice that T_1 depends only on δ. Moreover, by Lemma 4.3 (inequality (4.8_1)) we have [(1)]

$$\|\boldsymbol{v}(T_1)\|_{m+2} \le C\|\boldsymbol{v}_0\|_{m+2}$$

for C independent of $\|\boldsymbol{v}_0\|_{m+2}$ and T_1. Choosing $\mathcal{D} \le \frac{\delta}{C} = \delta_0$ we get the existence on $[T_1, 2T_1]$ and again by Lemma 4.2

$$\|\boldsymbol{v}(2T_1)\|_{m+2} \le C\,\mathcal{D} \le \varepsilon_0\,.$$

Repeating this procedure we obtain the solution on $[0, T)$ which satisfies properties (4.16).

5. Existence and Uniqueness for the $\mathcal{BVP}$.

In this section we shall show that the $\mathcal{BVP}$ (1.14), (1.15) admits a unique classical solution. For this result to hold, *we don't need any assumption on $|\alpha_1 + \alpha_2|$ or on the signof α_1 which can thus be any real number.*

These results will be obtained by means of a fixed point procedure that we are about to describe. Applying the Helmholtz decomposition (see Lemma 2.2) to $\Delta\boldsymbol{v}$, and taking into account (1.7), (1.13) we have

$$\left.\begin{aligned} &\Delta\boldsymbol{v} = \boldsymbol{u} + \operatorname{grad}\pi \\ &\operatorname{div}\boldsymbol{v} = 0 \end{aligned}\right\} \text{ in } \Omega$$

$$\boldsymbol{v} = 0 \text{ at } \partial\Omega$$

where, in view of (1.14), the field $\boldsymbol{u}$ satisfies

$$\left.\begin{aligned} &\nu\boldsymbol{u} + \alpha_1 \operatorname{curl}\boldsymbol{u} \times \boldsymbol{v} - \boldsymbol{\omega} \times \boldsymbol{v} = -\operatorname{grad} p + \boldsymbol{N}(\boldsymbol{v}) + \boldsymbol{F} \\ &\operatorname{div}\boldsymbol{u} = 0 \end{aligned}\right\} \text{ in } \Omega \tag{5.1}$$

$$\boldsymbol{u}\cdot\boldsymbol{n} = 0 \text{ at } \partial\Omega$$

(1) It is clear that $\|\boldsymbol{v}(T)\|_{m+2}$ is finite, since $\boldsymbol{v} \in C(0,T;\boldsymbol{X}_{m+1}) \cap L^\infty(0,T_1;\boldsymbol{H}^{m+2}(\Omega))$.

with $\boldsymbol{\omega}$ and $\boldsymbol{N}(\boldsymbol{v})$ given in (1.9) and (3.2), respectively, and $p = P + \nu\pi$. Take the Banach space

$$Y = \boldsymbol{X}_{m-1}, \quad m \geq 1$$

and, for $D > 0$ define

$$G = \left\{\boldsymbol{\varphi} \in Y: \ \|\boldsymbol{\varphi}\|_m \leq D, \right\}.$$

Consider now the map Φ defined in G, in the following way:

$$\Phi: \boldsymbol{\varphi} \mapsto \boldsymbol{u}$$

is the composition of the operator $\boldsymbol{\varphi} \mapsto \boldsymbol{v}$ defined by

$$\left.\begin{array}{l} \Delta \boldsymbol{v} = \boldsymbol{\varphi} + \operatorname{grad} \pi \\ \operatorname{div} \boldsymbol{v} = 0 \end{array}\right\} \quad \text{in } \Omega \tag{5.2}$$

$$\boldsymbol{v} = 0 \quad \text{at } \partial\Omega$$

with the operator $\boldsymbol{v} \mapsto \boldsymbol{u}$, defined by (5.1). To prove the existence of a solution to our original problem (1.14), (1.15) is equivalent to show that the map

$$\Phi: \ G \subset Y \ \rightarrow \ Y$$

admits a fixed point. The existence of such a fixed point will be proved by means of the Schauder theorem stated in Lemma 4.1. Specifically, we have

Theorem 5.1. *Let $\boldsymbol{F} \in \boldsymbol{H}^m(\Omega)$, $m \geq 2$. Then, there exists a positive constant $c = c(\Omega, m, |\alpha_1 + \alpha_2|, \alpha_1, \nu)$ such that if*

$$\|\boldsymbol{F}\|_m \leq c,$$

problem (1.14), (1.15) admits a unique solution

$$\boldsymbol{v} \in \boldsymbol{X}_{m+2}, \quad P \in \boldsymbol{H}^{m+1}(\Omega).$$

In particular, if $m = 3$, then

$$\boldsymbol{v} \in C^3(\overline{\Omega}), \quad P \in C^1(\overline{\Omega})$$

and the solution is classical.

Proof. From Lemma 3.4 we know that the map Φ is well-defined and that, if D is suitably restricted, *i.e.*, $\|\boldsymbol{F}\|_m$ is sufficiently small, $\Phi(G) \subset G$. Moreover, the compact embedding $\boldsymbol{X}_m \hookrightarrow \boldsymbol{X}_{m-1}$ implies the compactness of Φ. Then, we have only to check the continuity of Φ in the Y-norm. To this end, we observe that it is sufficient to show the continuity in L^2-norm. In fact, if $\Phi(\boldsymbol{\varphi}_n) \rightarrow \Phi(\boldsymbol{\varphi})$, in L^2 when $\boldsymbol{\varphi}_n \rightarrow \boldsymbol{\varphi}$ in $\boldsymbol{X}_{m-1}$, $\{\boldsymbol{\varphi}_n\}_{n\in\mathbb{N}}$, by the compactness of $\Phi(G)$ in $\boldsymbol{X}_{m-1}$ we find

$\Phi(\varphi_{n_k}) \to \boldsymbol{U}$ in $\boldsymbol{X}_{m-1}$, along a subsequence. However, by the convergence in L^2, we infer $\boldsymbol{U} = \Phi(\varphi)$ and since $\Phi(\varphi)$ is uniquely determined, the *whole* sequence $\Phi(\varphi_n)$ converges to $\Phi(\varphi)$ in $\boldsymbol{X}_{m-1}$. It remains to show convergence in L^2. In this regard, we denote by $\boldsymbol{v}_n$ and $\boldsymbol{v}$ the solutions to (5.2) corresponding to φ_n and φ, respectively, and set $\boldsymbol{u}_n = \Phi(\varphi_n)$, $\boldsymbol{u} = \Phi(\varphi)$. We now take the difference between equation (5.1_1) written for $\boldsymbol{u}$ and the same equation written for $\boldsymbol{u}_n$, and multiply the resulting equation by $\boldsymbol{u} - \boldsymbol{u}_n$. Integrating by parts over Ω and using $(5.1_{2,3})$ we find

$$\begin{aligned} \nu\|\boldsymbol{u} - \boldsymbol{u}_n\|^2 &+ \alpha_1(\operatorname{curl}(\boldsymbol{u} - \boldsymbol{u}_n) \times \boldsymbol{v}, \boldsymbol{u} - \boldsymbol{u}_n) \\ &+\alpha_1(\operatorname{curl}\boldsymbol{u}_n \times (\boldsymbol{v} - \boldsymbol{v}_n), \boldsymbol{u} - \boldsymbol{u}_n) - ((\boldsymbol{\omega} - \boldsymbol{\omega}_n) \times \boldsymbol{v}, \boldsymbol{u} - \boldsymbol{u}_n) \\ &-(\boldsymbol{\omega}_n \times (\boldsymbol{v} - \boldsymbol{v}_n), \boldsymbol{u} - \boldsymbol{u}_n) = (\boldsymbol{N}(\boldsymbol{v}) - \boldsymbol{N}(\boldsymbol{v}_n), \boldsymbol{u} - \boldsymbol{u}_n) \end{aligned} \tag{5.3}$$

where $\boldsymbol{\omega}_n = \operatorname{curl}\boldsymbol{v}_n$. Using the Schwarz inequality along with Lemma 3.1, we obtain

$$\begin{aligned} |((\boldsymbol{\omega} - \boldsymbol{\omega}_n) \times \boldsymbol{v}, \boldsymbol{u} - \boldsymbol{u}_n)| &+ |(\boldsymbol{\omega}_n \times (\boldsymbol{v} - \boldsymbol{v}_n), \boldsymbol{u} - \boldsymbol{u}_n)| \\ &+|(\boldsymbol{N}(\boldsymbol{v}) - \boldsymbol{N}(\boldsymbol{v}_n), \boldsymbol{u} - \boldsymbol{u}_n)| \\ &\leq C\left(\|\boldsymbol{v}\|_{m+2} + \|\boldsymbol{v}_n\|_{m+2}\right)\|\boldsymbol{v} - \boldsymbol{v}_n\|_{m+2}\|\boldsymbol{u} - \boldsymbol{u}_n\|. \end{aligned} \tag{5.4}$$

Furthermore, using (3.4), the Schwarz inequality and the Sobolev embedding theorem, we deduce

$$\begin{aligned} (\operatorname{curl}(\boldsymbol{u} - \boldsymbol{u}_n) \times \boldsymbol{v}, \boldsymbol{u} - \boldsymbol{u}_n) &= ((\boldsymbol{u} - \boldsymbol{u}_n) \cdot \operatorname{grad}\boldsymbol{v}, \boldsymbol{u} - \boldsymbol{u}_n) \\ &+ ((\boldsymbol{u} - \boldsymbol{u}_n) \times \operatorname{curl}\boldsymbol{v}, \boldsymbol{u} - \boldsymbol{u}_n) \\ &\leq C\|\boldsymbol{v}\|_{m+2}\|\boldsymbol{u}\|^2 \end{aligned} \tag{5.5}$$

Taking into account that

$$\|\boldsymbol{v}_n\|, \|\boldsymbol{v}\| \leq D, \quad \|\boldsymbol{v} - \boldsymbol{v}_n\|_{m+2} \leq C\|\varphi - \varphi_n\|_m$$

from (5.3)-(5.5) we conclude

$$\nu\|\boldsymbol{u} - \boldsymbol{u}_n\| \leq C_1\|\varphi - \varphi_n\|_m,$$

which, by what we said, is enough to ensure the continuity of the map Φ. The existence part is thus completed. Concerning uniqueness, we let $\boldsymbol{v}_1, \boldsymbol{v}_2$ be two solutions corresponding to the same data and denote by $\boldsymbol{u}_1, \boldsymbol{u}_2$ the projection of $\Delta\boldsymbol{v}_1$ and $\Delta\boldsymbol{v}_2$, respectively, on the space $\boldsymbol{X}_0$. We recall that $\boldsymbol{u}_i$, $i = 1, 2$ satisfies a problem of the type (5.1). Letting $\boldsymbol{u} = \boldsymbol{u}_1 - \boldsymbol{u}_2$ and using a procedure similar to that used to prove the continuity of the map Φ, we can show

$$\nu\|\boldsymbol{u}\|_m \leq C(\|\boldsymbol{v}_1\|_{m+2} + \|\boldsymbol{v}_2\|_{m+2})\|\boldsymbol{u}\|_m.$$

Therefore, since

$$\|v_1\|_{m+2} + \|v_2\|_{m+2} \le C_2 D = C_3 \|F\|_m,$$

uniqueness is secured whenever $\|F\|_m$ is sufficiently small. The proof of the theorem is completed.

6. Stability of Steady-State Solutions.

The aim of this section is to investigate the stability properties of the classical solutions to problem (1.14), (1.15), whose existence we proved in Section 5. Let v_0, P_0 be a solution to (1.14), (1.15) corresponding to a given body force F. Assume that at a given instant, say $t = 0$, v_0 is varied by a certain amount $w = w(x)$. The fluid will then undergo a new motion $v = v_0 + w(x,t)$, $P = P_0 + \pi(x,t)$ which will be a solution to the unsteady problem (1.8)-(1.11), (1.13) corresponding to the same body force F and with initial condition $v(x,0) = v_0 + w(x,0)$. The stability properties of the *basic flow* v_0, P_0 will depend on the behaviour in time of the *perturbation* $w(x,t), P(x,t)$. Our objective is to determine conditions under which the basic flow will be stable. First of all, we wish to point out that in order to perform this study, we should require $\alpha_1 \ge$. Actually, as we shall show, if such a condition is violated, the rest state, *i.e.*, $v_0 \equiv 0, P_0 \equiv$ constant, is *unstable*, a situation, this latter, which is unacceptable from the physical point of view. To show this instability property assume $\alpha_1 < 0$ and define a norm through

$$[\![v]\!] \equiv \max_{x\in\Omega} |\mathrm{grad}\, v(x)| + \|D^2 v\|.$$

Let us suppose for contradiction that the rest state is stable in this norm. Then

$$\forall \varepsilon > 0,\ \exists \delta(\varepsilon) > 0 : [\![v(0))]\!] < \delta \Longrightarrow \sup_{t\ge 0} [\![v(t)]\!] < \varepsilon. \tag{6.1}$$

With the definition

$$N(t) \equiv \tfrac{1}{2} \int_\Omega \left(|\alpha_1|\, |\mathrm{grad}\, v|^2 - |v|^2 \right),$$

multiplying (1.8) by v and integrating by parts over Ω we find

$$\frac{dN}{dt} = \nu \|\mathrm{grad}\, v\|^2 + (N(v), v) \tag{6.2}$$

with $N(v)$ defined in (3.2). By a simple calculation one can show

$$(N(v), v) = 4(\alpha_1 + \alpha_2) \int_\Omega [\mathrm{grad}\, v (\mathrm{grad}\, v)^T] \mathrm{grad}\, v \ge -4|\alpha_1 + \alpha_2| [\![v]\!] \|\mathrm{grad}\, v\|^2.$$

Thus, from (6.1) and (6.2) it follows that we can choose ε so small as to obtain

$$\frac{dN}{dt} \ge \lambda N(t), \quad \lambda := \frac{\nu - 4\varepsilon|\alpha_1 + \alpha_2|}{|\alpha_1|} > 0.$$

As a consequence,

$$N(t) \geq N(0)\exp(\lambda t). \tag{6.3}$$

Now, let $\boldsymbol{u}_n$ be the normalized eigenfunctions of the Stokes operator in Ω, *i.e.*,

$$\left.\begin{aligned} -\Delta \boldsymbol{u}_n &= \mu_n \boldsymbol{u}_n + \operatorname{grad} p \\ \operatorname{div} \boldsymbol{u}_n &= 0 \end{aligned}\right\} \text{ in } \Omega$$

$$\boldsymbol{u}_n = 0 \text{ at } \partial\Omega.$$

It is well-known that $\mu_n \to \infty$ as $n \to \infty$. We choose as initial data $\boldsymbol{v}(0) = A\boldsymbol{u}_n$ for sufficiently large n and suitable $A \in \mathbb{R}$ such that

$$\|\boldsymbol{v}(0)\| = A\|\boldsymbol{u}_n\| < \delta,$$

$$\frac{\|\operatorname{grad} \boldsymbol{v}(0)\|^2}{\|\boldsymbol{v}(0)\|^2} = \mu_n > 1/|\alpha_1|.$$

It follows from (6.3) that

$$\|\boldsymbol{v}(t)\|^2 \geq \frac{(|\alpha_1|\mu_n - 1)}{|\alpha_1|}\|\boldsymbol{v}(0)\|_2 \exp(\lambda t)$$

and hence

$$\|\boldsymbol{v}(t)\| \to \infty,$$

contradicting the hypothesis of stability.

These arguments push us to assume non-negative α_1. *In the sequel, we shall take* $\alpha_1 > 0$.

Coming back to the general stability problem, we introduce the non-dimensional quantities

$$\lambda = \frac{V\sqrt{\alpha_1}}{\nu}, \quad \sigma = \frac{V}{\nu\sqrt{\alpha_1}}$$

with V a characteristic velocity of the flow. Therefore, from (1.8)-(1.11), (1.13) we obtain that the perturbation $\boldsymbol{w}, p$ obeys the following initial-value problem

$$\left.\begin{aligned} &\frac{\partial}{\partial t}(\boldsymbol{w} - \Delta\boldsymbol{w}) - \Delta\boldsymbol{w} + \lambda[(\boldsymbol{\omega} - \Delta\boldsymbol{\omega}) \times \boldsymbol{v}_0 \\ &\qquad + (\boldsymbol{\omega} - \Delta\boldsymbol{\omega}) \times \boldsymbol{w} + (\boldsymbol{\omega}_0 - \Delta\boldsymbol{\omega}_0) \times \boldsymbol{w}] \\ &\qquad = -\operatorname{grad} p + \sigma \boldsymbol{N}_1(\boldsymbol{w}) \\ &\operatorname{div} \boldsymbol{w} = 0 \end{aligned}\right\} \text{ in } \Omega \times [0,\infty) \tag{6.4}$$

$$\boldsymbol{w}(x,t) = 0 \quad \text{at } \partial\Omega \times [0,\infty)$$

$$\boldsymbol{w}(x,0) = \boldsymbol{w}_0 \quad x \in \Omega$$

where

$$\omega_0 = \operatorname{curl} \boldsymbol{v}_0, \quad \boldsymbol{N}_1(\boldsymbol{w}) = \boldsymbol{N}(\boldsymbol{v}_0 + \boldsymbol{w}) - \boldsymbol{N}(\boldsymbol{w}).$$

Our next objective is to derive two independent suitable "energy equations" for $\boldsymbol{w}$. To reach this goal, we shall make the assumption that Ω is simply connected, so that we can employ Lemma 2.3 several times. Let us multiply (6.4_1) by $\boldsymbol{w}$ and integrate over Ω to obtain:

$$\tfrac{1}{2}\frac{d}{dt}\|\boldsymbol{w}\|_1^2 + \|\operatorname{grad} \boldsymbol{w}\|^2 + \lambda \int_\Omega (\omega - \Delta\omega) \times \boldsymbol{v}_0 \cdot \boldsymbol{w} = \sigma(\boldsymbol{N}_1(\boldsymbol{w}), \boldsymbol{w}).$$

Next, putting

$$\boldsymbol{u} = \operatorname{curl}(\boldsymbol{w} - \Delta\boldsymbol{w}), \quad \boldsymbol{u}_0 = \operatorname{curl}(\boldsymbol{v}_0 - \Delta\boldsymbol{v}_0), \tag{6.5}$$

we observe that the preceding identity can be written as

$$\tfrac{1}{2}\frac{d}{dt}\|\boldsymbol{w}\|_1^2 + \|\operatorname{grad} \boldsymbol{w}\|^2 + \lambda(\boldsymbol{u} \times \boldsymbol{v}_0, \boldsymbol{w}) = \sigma(\boldsymbol{N}_1(\boldsymbol{w}), \boldsymbol{w}). \tag{6.6}$$

Moreover, from (6.4_1) it follows that

$$\frac{\partial \boldsymbol{u}}{\partial t} + \boldsymbol{u} - \operatorname{curl} \boldsymbol{w} + \lambda\{\operatorname{curl}(\boldsymbol{u} \times \boldsymbol{v}_0) + \operatorname{curl}(\boldsymbol{u} \times \boldsymbol{w}) + \operatorname{curl}(\boldsymbol{u}_0 \times \boldsymbol{w})\} = \sigma \operatorname{curl} \boldsymbol{N}_1(\boldsymbol{w}).$$

Multiplying this equation by $\boldsymbol{u}$, integrating over Ω and taking into account that $\boldsymbol{w}$ is solenoidal, we find

$$\begin{aligned}\tfrac{1}{2}\frac{d}{dt}\|\boldsymbol{u}\|^2 + \|\boldsymbol{u}\|^2 = &(\operatorname{curl} \boldsymbol{w}, \boldsymbol{u}) + \lambda\{(\operatorname{curl}(\boldsymbol{u} \times \boldsymbol{v}_0), \boldsymbol{u}) \\ &+(\operatorname{curl}(\boldsymbol{u} \times \boldsymbol{w}), \boldsymbol{u}) + (\operatorname{curl}(\boldsymbol{u}_0 \times \boldsymbol{w}), \boldsymbol{u})\} \\ &+\sigma(\operatorname{curl} \boldsymbol{N}_1(\boldsymbol{w}), \boldsymbol{w})\end{aligned} \tag{6.7}$$

The "energy identities" (6.6), (6.7) are the starting point of our stability analysis. The next goal is to increase suitably the right-hand sides of these identities. To this end, we introduce the following notation

$$\overline{\boldsymbol{g}} = \max_{x \in \Omega} |\boldsymbol{g}(x)|.$$

We have

$$|(\boldsymbol{u} \times \boldsymbol{v}_0, \boldsymbol{w})| \le \overline{\boldsymbol{v}_0}\|\boldsymbol{u}\|\|\boldsymbol{w}\|. \tag{6.8}$$

Moreover, using Lemmas 2.4 and 3.1 it is not hard to show that

$$\begin{aligned}|(\boldsymbol{N}_1(\boldsymbol{w}), \boldsymbol{w})| &\le \|\boldsymbol{N}_1(\boldsymbol{w})\|\|\boldsymbol{w}\| \le c(\|\boldsymbol{v}_0\|_3 + \|\boldsymbol{w}\|_3)\|\boldsymbol{w}\|_3\|\boldsymbol{w}\| \\ &\le c(\|\boldsymbol{v}_0\|\|\boldsymbol{u}\|\|\boldsymbol{w}\| + \|\boldsymbol{u}\|^2\|\boldsymbol{w}\|)\end{aligned} \tag{6.9}$$

Substitution of (6.8) and (6.9) in (6.6) furnishes:

$$\frac{1}{2}\frac{d}{dt}\|\boldsymbol{w}\|_1^2 + \|\mathrm{grad}\,\boldsymbol{w}\|^2 + \lambda\overline{v}_0 + c|\sigma|\|\boldsymbol{v}_0\|_3)\|\boldsymbol{u}\|\|\boldsymbol{w}\| + |\sigma|\|\boldsymbol{u}\|^2\|\boldsymbol{w}\|. \tag{6.10}$$

Now, using the inequalities:

$$\|\mathrm{grad}\,\boldsymbol{w}\| \geq \lambda_1\|\boldsymbol{w}\|_1,\ \lambda_1 = \lambda_1(\Omega); \quad \|\boldsymbol{w}\|_0 \leq \|\boldsymbol{w}\|_1 \tag{6.11}$$

from (6.10) it follows that

$$\frac{d}{dt}\|\boldsymbol{w}\|_1 + \lambda_1\|\nabla\boldsymbol{w}\| \leq (\lambda\overline{v}_0 + c|\sigma|\|\boldsymbol{v}_0\|_3)\|\boldsymbol{u}\|_0 + c|\sigma|\|\boldsymbol{u}\|^2. \tag{6.12}$$

Concerning equation (6.7), the use of the Schwarz inequality, of the result of Lemmas 2.3 and 3.1, and of simple vector analysis formulas yields

$$\begin{aligned}
&|(\mathrm{curl}\,\boldsymbol{w},\boldsymbol{u})| \leq \|\mathrm{grad}\,\boldsymbol{w}\|\|\boldsymbol{u}\| \\
&|(\mathrm{curl}\,(\boldsymbol{u}\times\boldsymbol{v}_0,\boldsymbol{u})| = |(\boldsymbol{u}\cdot\mathrm{grad}\,\boldsymbol{v}_0,\boldsymbol{u})| \leq \overline{\mathrm{grad}\,\boldsymbol{v}_0}\|\boldsymbol{u}\|^2 \\
&|(\mathrm{curl}\,(\boldsymbol{u}\times\boldsymbol{w}),\boldsymbol{u})| = |(\boldsymbol{u}\cdot\mathrm{grad}\,\boldsymbol{w},\boldsymbol{u})| \leq \overline{\mathrm{grad}\,\boldsymbol{w}}\|\boldsymbol{u}\|^2 \\
&\qquad \leq c\|\boldsymbol{w}\|_3\|\boldsymbol{u}\|^2 \leq c\|\boldsymbol{u}\|^3 \\
&|(\mathrm{curl}\,(\boldsymbol{u}_0\times\boldsymbol{w}),\boldsymbol{u})| \leq |(\boldsymbol{w}\cdot\mathrm{grad}\,\boldsymbol{u}_0,\boldsymbol{u})| + |(\boldsymbol{u}_0\cdot\mathrm{grad}\,\boldsymbol{w},\boldsymbol{u})| \\
&\qquad \leq \overline{\mathrm{grad}\,\boldsymbol{u}_0}\|\boldsymbol{w}\|\|\boldsymbol{u}\| + \overline{\boldsymbol{u}_0}\|\boldsymbol{w}\|_1\|\boldsymbol{u}\|_0 \leq c(\overline{\mathrm{grad}\,\boldsymbol{u}_0} + \overline{\boldsymbol{u}_0})\|\boldsymbol{u}\|^2 \\
&|(\mathrm{curl}\,\boldsymbol{N}_1(\boldsymbol{w}),\boldsymbol{u})| \leq \|\mathrm{curl}\,\boldsymbol{N}_1(\boldsymbol{w})\|\|\boldsymbol{u}\| \leq c(\|\boldsymbol{v}\|_3\|\boldsymbol{u}\|^2 + \|\boldsymbol{u}\|^3).
\end{aligned} \tag{6.13}$$

Substituting these relations in (6.7), we obtain:

$$\begin{aligned}
\frac{d}{dt}\|\boldsymbol{u}\|^2 \leq &(\lambda\overline{\mathrm{grad}\,\boldsymbol{v}_0} + c\lambda(\overline{\mathrm{grad}\,\boldsymbol{u}_0} + \overline{\boldsymbol{u}_0}) + c|\sigma|\boldsymbol{v}_0\|_3 - 1)\boldsymbol{u}\|^2 \\
&+c(\lambda + |\sigma|)\|\boldsymbol{u}\|^3 + \|\mathrm{grad}\,\boldsymbol{w}\|\|\boldsymbol{u}\|,
\end{aligned}$$

from which

$$\frac{d}{dt}\|\boldsymbol{u}\| \leq -K\|\boldsymbol{u}\| + c(\lambda + |\sigma|)\|\boldsymbol{u}\|^2 + \|\mathrm{grad}\,\boldsymbol{w}\|, \tag{6.14}$$

where

$$K = -\lambda\overline{\mathrm{grad}\,\boldsymbol{v}_0} - c\lambda(\overline{\mathrm{grad}\,\boldsymbol{u}_0} + \overline{\boldsymbol{u}_0}) - c|\sigma|\|\boldsymbol{v}_0\|_3 + 1.$$

Multiplying (6.12) by $\tau > 0$ and adding it to (6.13), we get

$$\begin{aligned}
\frac{d}{dt}(\tau\|\boldsymbol{w}\|_1 + \|\boldsymbol{u}\|) + (\tau\gamma - 1)\|\mathrm{grad}\,\boldsymbol{w}\| + \{K - \tau(\lambda\boldsymbol{v}_0 + c|\sigma|\|\boldsymbol{v}_0\|_3)\}\|\boldsymbol{u}\| \\
c(\lambda + (1+\tau)|\sigma|)\|\boldsymbol{u}\|^2.
\end{aligned} \tag{6.15}$$

From here, it results that to find stability we have to require

$$\tau\lambda_1 - 1 > 0, \qquad K - \tau(\lambda\overline{v}_0 + c|\sigma|\|v_0\|_3) > 0. \tag{6.16}$$

In this way, setting

$$\tau' = \tau\lambda_1 - 1, \qquad K' = K - \tau(\lambda\overline{v}_0 + c|\sigma|\|v_0\|_3),$$

(6.14) becomes:

$$\frac{d}{dt}(\tau\|\boldsymbol{w}\|_1 + \|\boldsymbol{u}\|_0) + \tau'\|\text{grad}\,\mathbf{w}\| + K'\|\boldsymbol{u}\|_0 \leq C\|\boldsymbol{u}\|^2, \tag{6.17}$$

where $C = c(\lambda + (1+\tau)|\sigma|)$. Recalling ($6.11_1$), inequality (6.15) gives

$$\frac{d}{dt}(\tau\|\boldsymbol{w}\|_1 + \|\boldsymbol{u}\|) + \frac{\lambda_1\tau}{\tau'}\|\boldsymbol{w}\|_1 + K'\|\boldsymbol{u}\| \leq C\|\boldsymbol{u}\|^2. \tag{6.18}$$

Putting $\tau_0 = \min\{\lambda_1/\tau', K'\}$ and introducing the "energy"

$$E = \tau\|\boldsymbol{w}\|_1 + \|\boldsymbol{u}\|, \tag{6.19}$$

from (6.16) we have:

$$\frac{dE}{dt} + \tau_0 E \leq CE^2.$$

It is now a simple exercise to prove that, provided that $\tau > 0$ and that $E(0) < \tau_0/C$, there exists a positive quantity δ such that:

$$E(t) \leq E(0)\exp(-\delta t) \tag{6.20}$$

which,in turn, means asymptotic stability.

We have thus proved the following result

Theorem 6.1. *Let Ω be a bounded, simply connected domain of class C^3 and let $\alpha_1 > 0$. Moreover, let $s_0 = \{\boldsymbol{v}_0, P_0\}$ be a regular solution to problem* (1.17),(1.18). *Then, if conditions* (6.15) *are met for some $\tau > 0$, the flow s_0 is exponentially stable in the energy norm E defined by* (6.18), *provided E at time $t = 0$ is sufficiently small.*

Remark 6.1. From (6.15) it is clear that, for the stability region becomes smaller the larger is the value of $|\alpha_1 + \alpha_2|$. In fact, the condition $\alpha_1 + \alpha_2 = 0$ (suggested by the Clausius-Duhem inequality) acts in favour of stability.

References

[1] Adams R.A. (1975), *Sobolev spaces*, Academic Press, New York.

[2] Agmon, S., Douglis, A. & L. Nirenberg (1959) *Estimates near the boundary for solutions of elliptic partial differential equations satisfying general boundary conditions I*, Comm. Pure Appl. Math., **12**, 623-727.

[3] Cattabriga L. (1961), *Su un problema al contorno relativo al sistema di equazioni di Stokes* , Rend. Sem. Mat. Padova **31**, 308-340.

[4] Coscia V. & G.P.Galdi (1994) Existence, uniqueness and stability of regular steady motions of a second-grade fluid, Int. J. Non-Linear Mechanics, **29** (4), 493-506.

[5] Coleman B.D. & H. Markovitz (1964), *Normal stress effects in second-order fluids*, J. App. Phys. **35**, 1-48.

[6] Dunn J.E. & R.L. Fosdick (1974), *Thermodynamics, stability and boundedness of fluids of complexity* 2 *and fluids of second grade*, Arch. Rational Mech. Anal. **56**, 191-252.

[7] Dunn J.E. & K.R. Rajagopal (1994), *Fluids of differential type: critical review and thermodynamic analysis*, to appear in Int. Jl. Engng Sci.

[8] Fosdick R.L. & K.R. Rajagopal (1978), *Anomalous features in the model of second order fluids*, Arch. Rational Mech. Anal. **70**, 145-152.

[9] Foias C. & R. Temam (1978), *Remarques sur les équations de Navier-Stokes stationnaires et les phénomènes successifs de bifurcation*, Ann. Scuola Norm. Sup. Pisa, S. IV, **5**, 29-63 .

[10] Friedman A. (1969), *Partial differential equations*, Holt, Rinehart & Winston Inc., New York.

[11] Galdi G.P. (1994), *An introduction to the mathematical theory of the Navier-Stokes equations*, Vol. I: Linearized steady Problems, Springer Tracts in Natural Philosophy, Springer Verlag, Heidelberg.

[12] Galdi G.P. (1994), *An introduction to the mathematical theory of the Navier-Stokes equations*, Vol. II: Nonlinear steady problems, Springer Tracts in Natural Philosophy, Springer Verlag, Heidelberg.

[13] Galdi G.P., M. Grobbelaar-Van Dalsen & N. Sauer (1993), *Existence and uniqueness of classical solutions of the equations of motion for second grade fluids*, Arch. Rational Mech. Anal., **124**, 221-237.

[14] Galdi G.P., M. Grobbelaar-Van Dalsen & N. Sauer (1993), *Existence and Uniqueness of Solutions of the Equations of a Fluid of Second Grade with Non-homogenous Boundary Conditions*, to appear on Int. J. Non-Linear Mech.

[15] Galdi G.P., M. Padula & K.R. Rajagopal (1990), *On the conditional stability of the rest state of a fluid of second grade in unbounded domains*, Arch. Rational Mech. Anal. **109**, 173-182.

[16] Galdi G.P. & K.R. Rajagopal (1994), *On the slow of a body in a second-grade fluid*, to appear on Arch. Rational Mech. Anal.

[17] Galdi G.P. & A. Sequeira (1994), *Further existence results for classical solutions of the equations of a second-grade Fluid*, to appear on Arch. Rational Mech. Anal.

[18] Rajagopal K.R. (1992), *Flow of viscoelastic fluids between rotating plates*, Theor. and Comput. Fluid Dyn., **3**, 185-216.

[19] Rajagopal K.R. & P.N. Kaloni (1989), *Some remarks on boundary conditions for lows of fluids of the differential type*, in: Continuum Mechanics and its Applications, Hemisphere Press.

[20] Rivlin R.S. & J.L. Ericksen (1955), *Stress-deformation relations for isotropic materials*, J. Rational Mech. Anal. **4**, 323-425.

[21] Simader C.G. & H. Sohr (1993) *The weak and strong Dirichlet problem for Δ in L^q in bounded and exterior domains*, Pitman Research Notes in Mathematics, in press.

[22] Temam R. (1975), *On the Euler equation of incompressible perfect fluids*, J. Funct. Anal. **20**, 32-43.

[23] Temam R. (1986), *Remarks on the Euler equation*, In: Proc. Symp. Pure Math. **45**, 429-430.

[24] Truesdell C. & W. Noll (1965), *The nonlinear field theories of Mechanics*, Handbuch der Physik **III/3**, Springer-Verlag, Heidelberg.

RECENT RESULTS ON WAVE PROPAGATION IN CONTINUUM MODELS

T. Ruggeri
C.I.R.A.M. - University of Bologna, Bologna, Italy

I. Hyperbolic Balance Laws Systems and Wave Propagation

In this Chapter we present some mathematical definitions and some results of non linear wave problems for a generic quasi-linear hyperbolic system of balance laws type.

1 Hyperbolic Systems

Let us consider a generic first order quasi-linear system:

$$\mathbf{A}^\alpha(\mathbf{u})\partial_\alpha \mathbf{u} = \mathbf{f}(\mathbf{u}) \tag{1}$$

for the R^N unknown vector $\mathbf{u} \equiv \mathbf{u}(x^\alpha)$ $(\alpha = 0, i;\ i = 1,2,3;\ x^o = t;\ \partial_\alpha \equiv \partial/\partial x^\alpha)$. The matrices $\mathbf{A}^\alpha$ $(N \times N)$ and the column R^N vector of the productions $\mathbf{f}$ are functions of the field $\mathbf{u}$.
The system (1) becomes semi-linear if the matrices $\mathbf{A}^\alpha$ are constants, and linear if the $\mathbf{A}^\alpha$ are constant and $\mathbf{f}$ is a linear function of $\mathbf{u}$.

Def. 1.1 (Balance Law Systems) *A balance law system is a particular case of* (1), *when there exists a choice of the field* $\mathbf{u}$ *such that:*

$$\mathbf{A}^\alpha = \frac{\partial \mathbf{F}^\alpha}{\partial \mathbf{u}}. \tag{2}$$

In this case (1) *becomes*

$$\partial_\alpha \mathbf{F}^\alpha(\mathbf{u}) = \mathbf{f}(\mathbf{u}). \tag{3}$$

We give now some definitions relative to the hyperbolicity.

Def. 1.2 (Hyperbolicity in the t-direction) *The system* (1) *is said to be hyperbolic in the t-direction if*

- $\det \mathbf{A}^o \neq 0$;
- *the eigenvalue problem*

$$(\mathbf{A}_n - \lambda \mathbf{A}^o)\,\mathbf{d} = 0; \quad \mathbf{A}_n = \mathbf{A}^i n_i \tag{4}$$

admits $\forall \mathbf{n} \in R^3$ *only real eigenvalues* λ *and a set of linearly independent right eigenvectors* $\mathbf{d}$. *This implies in particular that if an eigenvalue has multiplicity* m *there exist* m *linear independent eigenvectors.*

The λ*'s are called characteristic velocities and the polynomial associated to* (4) *is the characteristic polynomial.*

In the same way it is possible to define the left eigenvectors $\mathbf{l}$

$$\mathbf{l}\,(\mathbf{A}_n - \lambda \mathbf{A}^o) = 0. \tag{5}$$

It is well known from linear algebra that it is possible always to choose the left and right eigenvectors such that:

$$\mathbf{l}^{(i)} \cdot \mathbf{d}^{(j)} = \delta^{ij}, \quad \forall i, j = 1, 2, \ldots N. \tag{6}$$

Def. 1.3 (Strictly hyperbolicity) *If all the eigenvalues* λ *are distinct:*

$$\lambda^{(1)} < \lambda^{(2)} < \ldots < \lambda^{(N)} \tag{7}$$

the hyperbolic system is called a strictly hyperbolic system.

Def. 1.4 (Genuine non-linearity) *An hyperbolic system* (1) *is said to be genuine non-linear if for each eigenvalues* λ*'s we have:*

$$\frac{\partial \lambda^{(i)}}{\partial \mathbf{u}} \cdot \mathbf{d}^{(i)} \neq 0, \quad \forall i = 1, 2, \ldots, N; \quad \forall \mathbf{u} \in \mathcal{D} \subseteq \mathcal{R}^N. \tag{8}$$

In contrast we have the so called exceptionality condition:

Def. 1.5 (Exceptional waves) *A characteristic velocity (and the corresponding wave) is said to be exceptional, if*

$$\frac{\partial \lambda}{\partial \mathbf{u}} \cdot \mathbf{d} \equiv 0, \quad \forall \mathbf{u} \in \mathcal{D} \subseteq \mathcal{R}^N. \tag{9}$$

If the previous condition is satisfied for all the eigenvalues, the hyperbolic system is

Nevertheless, exceptional eigenvalues are very common for the systems of Mathematical Physics and they play a relevant role in several questions of non linear wave propagation, in particular for systems in more of one-space dimension [1]. A typical example is the contact velocity in fluid dynamics ($\lambda = \mathbf{v} \cdot \mathbf{n}$, where $\mathbf{v}$ is the fluid velocity). The occurrence of exceptional waves in mathematical physics appears clear, if one takes into account the following theorem [2]:

Theorem 1.1 *All the characteristic velocities λ of a system of balance laws that are multiple eigenvalues of the characteristic polynomial are exceptionals.*

For example in the fluid case the contact wave has multiplicity 3.

Def. 1.6 (Symmetric Systems) *A system* (1) *is said Symmetric Hyperbolic or briefly Symmetric (Friedrich's definition) if*

- $\mathbf{A}^\alpha = (\mathbf{A}^\alpha)^T$; *i.e.* $\mathbf{A}^\alpha$ *are all symmetric;*
- $\mathbf{A}^o$ *is positive definite.*

From linear algebra each Symmetric System is automatically hyperbolic; it is not true the viceversa, i.e. the class of Symmetric Hyperbolic Systems is contained in the class of Hyperbolic Systems.

1.1 Entropy Principle and Symmetric Systems

For a system of balance laws (3), one of the most important selection rule for the admissible constitutive equations is the *entropy principle.* At the present level of generality, this principle requires that each solutions of (2) must satisfy the supplementar scalar inequality:

$$\partial_\alpha h^\alpha(\mathbf{u}) = g(\mathbf{u}) \leq 0, \tag{10}$$

where $-h^o$ and $-h^i$ represent the entropy density and the entropy flux respectively. The problem of exploiting this principle was considered by different authors (Godunov 1961 [3], Friedrichs & Lax 1971 [4], Boillat 1974 [5], Ruggeri & Strumia 1981 [6], Ruggeri [7] 1985). In particular in [5]-[7] the following theorem was proved:

Theorem 1.2 *There exists a* **main field** $\mathbf{u}'$ *and* 4 **potentials** h'^α *such that*

$$\mathbf{F}^\alpha = \frac{\partial h'^\alpha}{\partial \mathbf{u}'}, \tag{11}$$

where

$$h^\alpha = \mathbf{u}' \cdot \mathbf{F}^\alpha - h'^\alpha; \qquad \mathbf{u}' \cdot \mathbf{f} \leq 0. \tag{12}$$

Inserting (11), the system (1) becomes

$$\mathbf{A}'^{\alpha}\partial_{\alpha}\mathbf{u}' = \mathbf{f}(\mathbf{u}')$$

where

$$\mathbf{A}'^{\alpha} = \frac{\partial^2 h'^{\alpha}}{\partial \mathbf{u}' \partial \mathbf{u}'}.$$

For this reason if $\mathbf{A}'^{o}$ is positive definite the system is Symmetric Hyperbolic. Therefore, in order to have the system in a symmetric form it is necessary that the following quadratic form is positive:

$$Q = \delta\mathbf{u}' \cdot \mathbf{A}'^{o} \cdot \delta\mathbf{u}' = \delta\mathbf{u}' \cdot \delta\mathbf{F}^{o} = \delta\mathbf{u}' \cdot \frac{\partial^2 h'^{o}}{\partial \mathbf{u}' \partial \mathbf{u}'} \cdot \delta\mathbf{u}' > 0 \quad \forall \delta\mathbf{u}'. \tag{13}$$

We call this mathematical condition the *Symmetrizability Condition* and taking into account the last expression in (13) we note that the symmetrizability corresponds to the convexity of h'^{o} with respect to the main field components $\mathbf{u}'$. Taking into account $(12)_1$ we observe that h'^{o} is the Legendre transformation of h^{o} and $\mathbf{u}'$ and $\mathbf{u} \equiv \mathbf{F}^{o}$ are conjugate variables:

$$\mathbf{u}' = \frac{\partial h^{o}}{\partial \mathbf{u}}; \qquad \mathbf{u} = \frac{\partial h'^{o}}{\partial \mathbf{u}'},$$

and then $Q = \delta^2 h^{o}$. So in the classical case the symmetrizability condition is equivalent to the physical assumption of convexity. On the other hand the convexity of $-h^{o}$, i.e. the concavity of entropy density, is in the thermodynamical theories a natural request of thermodynamical stability. Then we conclude that thermodynamical assumptions imply as consequence the symmetric form of the system and for well known theorems the Cauchy problem (locally in time) is well posed for smooth initial data. We quote in particular the theorem of Fischer and Marsden:

Theorem 1.3 *Each symmetric system, whose the initial data are of class $H^s(R^N)$ with $s \geq 4$, has unique solution $\in H^s(R^N)$ in the neighbourhood of the initial manifold, even if the system is not strictly hyperbolic.*

2 Linear Waves

First of all we give some general results for the usual linear plane harmonic waves associated to the linearized version of the system (1). We consider for simplicity the case of one space dimension and we rewrite the non linear system (1) putting $\mathbf{A}^{o} \equiv \mathbf{I}$ without loss of generality:

$$\mathbf{u}_t + \mathbf{A}(\mathbf{u})\mathbf{u}_x = \mathbf{f}(\mathbf{u}). \tag{14}$$

Now linearize the system (14) in a neighbourhood of a constant state $\mathbf{u}_o$ (that is a solution of (14) if $\mathbf{f}(\mathbf{u}_o) = 0$):

$$\mathbf{u} = \mathbf{u}_o + \tilde{\mathbf{u}} \tag{15}$$

where $\tilde{\mathbf{u}}$ is a small perturbation. One obtains

$$\tilde{\mathbf{u}}_t + \mathbf{A}_o \tilde{\mathbf{u}}_x = \mathbf{B}_o \tilde{\mathbf{u}} \tag{16}$$

where $\mathbf{A}_o \equiv \mathbf{A}(\mathbf{u}_o)$, $\mathbf{B}_o \equiv (\nabla \mathbf{f})_o$.
Then let us look for solutions of (16) of the kind

$$\tilde{\mathbf{u}} = \mathbf{w} e^{i(\omega t - \mathcal{K} x)} \tag{17}$$

representing a plane harmonic wave of frequency $\omega \in \Re^+$ and complex wave number $\mathcal{K} = \mathcal{K}_r + i\mathcal{K}_i$, propagating in the x-direction ($\mathcal{K}_r = \mathcal{R}e(\mathcal{K})$, $\mathcal{K}_i = \mathcal{I}m(\mathcal{K})$ and $\mathbf{w} = \text{const.} \in C$).
By substituting (17) in (16) one has the homogeneous linear system

$$\left(\mathbf{I} - z\mathbf{A}_o + \frac{i}{\omega}\mathbf{B}_o\right)\mathbf{w} = 0 \tag{18}$$

where $z = \mathcal{K}/\omega$. From (18) we obtain the dispersion relation

$$\det\left(\mathbf{I} - z\mathbf{A}_o + \frac{i}{\omega}\mathbf{B}_o\right) = 0, \tag{19}$$

which enables us to get the phase velocity v_{ph} and the attenuation factor α in terms of the frequence parameter ω

$$v_{ph} = \frac{\omega}{\mathcal{R}e(\mathcal{K})} = \frac{1}{\mathcal{R}e(z)}; \qquad \alpha = -\mathcal{I}m(\mathcal{K}) = -\omega \mathcal{I}m(z). \tag{20}$$

For satisfying the linear stability, $\alpha(\omega)$ must be positive (negative) for the waves travelling in the $x > 0$ (< 0) region, i.e.

$$\alpha(\omega)x > 0. \tag{21}$$

Since the characteristic line C starting from $x = 0$ is $x = \lambda_o t$, the previous condition (21) becomes

$$\alpha(\omega)\lambda_o > 0, \qquad \forall \omega \in R^+. \tag{22}$$

2.1 High frequency limit

First of all our aim is to estimate v_{ph} and α in the limit of high frequencies, i.e. when $\omega \to +\infty$. For this reason let us consider the formal development in series of power of the frequency parameter $1/\omega$:

$$z = \sum_{\alpha \geq 0} \frac{z_\alpha}{\omega^\alpha}, \qquad \mathbf{w} = \sum_{\beta \geq 0} \frac{\mathbf{w}_\beta}{\omega^\beta}. \tag{23}$$

Putting (23) in (18), with some calculations, the following recursive formula[(1)] for the $\alpha - th$ term in $1/\omega^\alpha$ is deduced (see [8])

$$(\mathbf{I} - z_o\mathbf{A})\mathbf{w}_\alpha - (z_1\mathbf{A} - i\mathbf{B})\mathbf{w}_{\alpha-1} - \sum_{\beta=2}^{\alpha} z_\beta \mathbf{A}\mathbf{w}_{\alpha-\beta} = 0 \qquad \alpha = 0, 1, 2, ... \tag{24}$$

where, by definition, $\mathbf{w}_\gamma = 0$ if $\gamma < 0$.
For $\alpha = 0$ one has from (24)

$$(\mathbf{I} - z_o\mathbf{A})\mathbf{w}_o = 0 \tag{25}$$

and so it follows

$$z_o = \frac{1}{\lambda}, \qquad \mathbf{w}_o = \mathbf{d} \tag{26}$$

where λ and $\mathbf{d}$ are respectively an eigenvalue and the corresponding right eigenvector of $\mathbf{A}$. Observe that z_o and $\mathbf{w}_o$ are real. For $\alpha = 1$ we have, recalling (26),

$$\left(\mathbf{I} - \frac{\mathbf{A}}{\lambda}\right)\mathbf{w}_1 - (z_1\mathbf{A} - i\mathbf{B})\mathbf{d} = 0. \tag{27}$$

Multiplying (27) by the left eigenvector $\mathbf{l}$ corresponding to λ we get

$$z_1 = \frac{i}{\lambda}(\mathbf{l} \cdot \mathbf{B} \cdot \mathbf{d}) \equiv i\frac{b_{\lambda\lambda}}{\lambda} \tag{28}$$

and so z_1 is a pure imaginary number. Hence at this stage

$$z = \frac{\mathcal{K}}{\omega} \sim z_o + \frac{z_1}{\omega} = \frac{1}{\lambda}\left[1 + \frac{i}{\omega}(\mathbf{l} \cdot \mathbf{B} \cdot \mathbf{d})\right] \qquad \text{for}\, \omega \gg 1 \tag{29}$$

and, from (20),

$$\lim_{\omega \to \infty} v_{ph}(\omega) = \lambda \tag{30}$$

$$\lim_{\omega \to \infty} \alpha(\omega) = -\frac{1}{\lambda}(\mathbf{l} \cdot \mathbf{B} \cdot \mathbf{d}). \tag{31}$$

$$\lim_{\omega \to \infty} \alpha(\omega)\lambda = -\mathbf{l} \cdot \mathbf{B} \cdot \mathbf{d}. \tag{32}$$

---

(1) In what follows we omit, for simplicity, the index o denoting the constant state.

The first furnishes the result that the phase velocity coincides, in the limit of high frequencies, with the characteristic velocity. The second and the last one give respectively the spatial and the temporal absorption. We summarize the previous results in this statement:

Statement 2.1 *In the limit of high frequencies, the phase velocity coincides with the characteristic velocity evaluated in the unperturbed state.*
The time absorption is given by (32) *and* $\mathbf{u}_o$ *is linearly stable with respect of high frequencies perturbation waves if and only if*

$$(\mathbf{l}\cdot\mathbf{B}\cdot\mathbf{d})_o < 0. \tag{33}$$

We shall see in the following that the previous linear stability implies a non linear stability, i.e. also the non linear discontinuity waves (acceleration waves) decay in the time if (33) is true.

3 Non linear waves

In this section we give a brief survey on two types of non liner waves for hyperbolic systems: discontinuity waves and shock waves.

3.1 Discontinuity Waves

For a system of type (1) it is possible to consider a particular class of solutions characterizing the so called *weak discontinuity waves* or, in the language of continuum mechanics, *acceleration waves*. Let us consider, in a space of generic dimension, a moving surface (*wave front*) Σ of cartesian equation $\varphi(\mathbf{x},t) = 0$ separating the space in two subspaces: ahead the wave front we have the known ***unperturbed field*** $\mathbf{u}_o(x,t)$ and behind it the unknown ***perturbed field*** $\mathbf{u}(\mathbf{x},t)$. Both the fields $\mathbf{u}_o$ and $\mathbf{u}$ are supposed regular solutions of (14) and continuous functions across the surface Σ, but with discontinuous normal derivative i.e.:

$$[\![\mathbf{u}]\!] = 0, \qquad [\![\mathbf{u}_\varphi]\!] = \mathbf{\Pi} \neq 0. \tag{34}$$

The square brackets indicate the *jump* (for simplicity we denote briefly with b and b_o the values of each quantity b evaluated on Σ respectively for $\varphi \to 0^-$ and $\varphi \to 0^+$)

$$[\![\cdot]\!] = (\cdot)_{\varphi=0^-} - (\cdot)_{\varphi=0^+}$$

and $\mathbf{u}_\varphi \equiv \partial\mathbf{u}/\partial\varphi$. The following results hold [1], [9], [10]:

1. The normal velocity $V = -\varphi_t/|\nabla\varphi|$ is equal to a *characteristic velocity* evaluated in $\mathbf{u}_o$

$$V = \lambda_o \equiv \lambda(\mathbf{u}_o). \tag{35}$$

2. The jump vector $\mathbf{\Pi}$ is proportional to the right eigenvector $\mathbf{d}$ of $\mathbf{A}$ (corresponding to the eigenvalue λ) evaluated in $\mathbf{u}_o$

$$\mathbf{\Pi} = \Pi \mathbf{d}_o \equiv \Pi \mathbf{d}(\mathbf{u}_o) \tag{36}$$

3. The *amplitude* Π satisfies a *Bernoulli equation*

$$\frac{d\Pi}{dt} + a(t)\Pi^2 + b(t)\Pi = 0 \tag{37}$$

where $d/dt = \partial_t + \lambda_o \partial_x$ indicates the time derivative along the characteristic lines $dx/dt = \lambda_o$ and $a(t)$, $b(t)$ are known functions of time through $\mathbf{u}_o$. In one space dimension, we have:

$$a(t) = \varphi_x (\nabla\lambda \cdot \mathbf{d})_o \, ; \tag{38}$$

$$b(t) = \left\{ \mathbf{d}^T \left((\nabla \S)^{\mathrm{T}} - \nabla \S \right) \cdot \frac{d\mathbf{u}}{dt} + (\nabla\lambda \cdot \mathbf{d})(\S \cdot \mathbf{u}_x)^{\cdot} - \nabla \mathbf{f} \cdot \mathbf{d} \right\}_o \, ; \tag{39}$$

$$\frac{d\varphi_x}{dt} + (\nabla\lambda \cdot \mathbf{u}_x)_o \varphi_x = 0 \, ; \qquad \varphi_x(0) = 1 \, . \tag{40}$$

Here $\nabla \equiv \partial/\partial \mathbf{u}$ and the index o denotes the quantities evaluated in $\mathbf{u}_o$. The solution of (37) is

$$\Pi(t) = \frac{\Pi(o) e^{-\int_o^t b(\xi)\, d\xi}}{1 + \Pi(o) \int_o^t a(\zeta) e^{-\int_o^\zeta b(\xi)\, d\xi} \, d\zeta} \, . \tag{41}$$

If the wave is *exceptional* (see (9)) then the Bernoulli equation becomes linear since the coefficient $a(t)$ vanishes. On the contrary, if the wave satisfies the *genuine non linearity* condition $\nabla\lambda \cdot \mathbf{d} \neq 0$ there exists, for a suitable initial amplitude $\Pi(0)$, a *critical time* such that the denominator of (41) tends to zero and the discontinuity becomes unbounded. This instant corresponds to the creation of a *strong discontinuity* (shock wave) and so the field itself presents a discontinuity across the wave front.

A qualitative analysis of the Bernoulli equation solution can be found in the papers [11], [10].

In what follows we assume that the genuine non linearity holds; therefore $a(t) \neq 0$ and, without loss of generality, we can suppose $a > 0$.

The problem of discontinuity waves is suitable to provide a measure of the non linear stability for the unperturbed field $\mathbf{u}_o$ in the sense of a definition given in [10]. In fact, we say that the regular solution $\mathbf{u}_o(\mathbf{x}, t)$ of (14) is λ-*stable* if a discontinuity wave propagating in the unperturbed field $\mathbf{u}_o$ has the amplitude $\Pi(t)$ which remains small ($\forall t \geq 0$) provided that the initial amplitude $\Pi(0)$ is sufficiently small. In particular we have *asymptotic λ-stability* if

$$\lim_{t\to\infty} |\Pi(t)| = 0. \tag{42}$$

In [10] it is shown that in terms of the Bernoulli equation coefficients $a(t)$, $b(t)$ the following integrability conditions are necessary and sufficient for the λ-stability of the solution $\mathbf{u_o}$

$$\int_0^\infty a(\xi)e^{-\int_0^\xi b(\zeta)\,d\zeta}\,d\xi = K(x_o) < +\infty \tag{43}$$

$$\int_0^t b(\xi)\,d\xi > m \qquad \forall t > 0 \tag{44}$$

where m is a constant. The condition (43) guarantees the existence of a finite threshold value $\Pi_{th} = 1/K > 0$ for which if

$$|\Pi(0)| < \Pi_{th}, \tag{45}$$

then the critical time never exists (i.e. $\Pi(t)$ exists $\forall t > 0$). Therefore (43) is a condition of global existence of $\Pi(t)$. The condition (44) implies that $|\Pi(t)|$ remains bounded for each time t. Moreover if, instead of (44), we have the stronger condition

$$\int_0^\infty b(t)\,dt = +\infty \tag{46}$$

then the solution is asymptotically λ-stable. In the case $\mathbf{u}_o$ = constant it is easily obtained, from (38)÷(40), that also a and b are constant and the λ-stability conditions (43), (44) are satisfied with the only condition $b > 0$. In the present case, observing that the constant solution exists if $\mathbf{f}(\mathbf{u}_o) = 0$, (39) becomes

$$b = -\mathbf{l}_o \cdot \mathbf{B}_o \cdot \mathbf{d}_o \tag{47}$$

with $\mathbf{B} = \nabla \mathbf{f}$. Therefore

$$b = -\mathbf{l}_o \cdot \mathbf{B}_o \cdot \mathbf{d}_o > 0 \tag{48}$$

is, in the non-linear case, the only condition for the λ-stability of a constant state; indeed we have asymptotic λ-stability because (46) is clearly verified. In the present case of constant solution $\Pi_{th} = b/a$ and the initial perturbation must satisfy

$$|\Pi(0)| < \frac{b}{a} \tag{49}$$

in order to have stability. We observe that in the non linear case if we consider solutions of *asymptotic waves* type [12], instead of the discontinuity waves, the same conditions (43), (44) for λ-stability are obtained, because the coefficients appearing in the transport equation have the same expression of the coefficients a, b in the

Bernoulli equation [13]. General connections between asymptotic waves and stability have been studied in the linear case in [14]. A possible extension to the non linear case was proposed in [15].
It is now easy to see that when the unperturbed state is constant then the linear stability implies (48), i.e. the non linear λ-stability.

3.1.1 Linear and Non linear stability

In fact supposing that the linear plane wave is stable, the condition (32) is true and it coincides with the λ–stability for the acceleration wave (48). Taking into account the previous considerations for linear waves and discontinuity waves we summarize:

Statement 3.1 *When the unperturbed state* $\mathbf{u}_o$ *is constant then:*

- *the phase velocity, in the limit of high frequency, coincides with the velocity of the discontinuity waves;*

- *the absorption factor* b *in the Bernoulli equation, controlling the amplitude decay of the discontinuity waves, coincides with the high frequency limit of the temporal absorption of linear plane waves and therefore if the wave is linearly stable with respect to plane waves propagation then it is also non linearly* λ*-asymptotic stable.*

3.2 Shock waves

If across the wave front the field $\mathbf{u}$ presents a jump we have a *shock wave.* Shock waves are possible only for systems of balance laws and, as it is well known, they are a particular class of the so called *weak solutions.* First of all we recall the definition of weak solution for systems of the type (3).

3.2.1 Weak solutions

Let us consider a domain C in the space-time of boundary Σ and we call ν^α the normal in Σ. At first let us multiply (3) by a generic scalar test function $\Phi(x^\beta)$ having support in C and integrate on C:

$$\int_C \Phi\left(\partial_\alpha \mathbf{F}^\alpha - \mathbf{f}\right) dC = 0. \tag{50}$$

Now, we rewrite (50) in the equivalent form:

$$\int_C \partial_\alpha\left(\Phi \mathbf{F}^\alpha\right) dC - \int_C \left(\mathbf{F}^\alpha \partial_\alpha \Phi + \mathbf{f}\Phi\right) dC = 0. \tag{51}$$

We apply then the Gauss-Green theorem, obtaining:

$$\int_{\Sigma} \nu_\alpha \Phi \mathbf{F}^\alpha \, d\Sigma - \int_{C} (\mathbf{F}^\alpha \partial_\alpha \Phi + \mathbf{f}\Phi) \, dC = 0. \tag{52}$$

The first integral vanishes and so we have:

$$\int_{C} (\mathbf{F}^\alpha \partial_\alpha \Phi + \mathbf{f}\Phi) \, dC = 0. \tag{53}$$

Def. 3.1 (Weak solutions) *A solution* $\mathbf{u}(x^\alpha)$ *is said a weak solution of* (3) *if it satisfies* (53) *for every test function* Φ *with support in* C.

The advantage of this kind of solution lies in the fact that now the differentiability conditions are not requested as for the classical solutions of (3). It is easily seen that each classical solution is also a weak solution, while the vice versa is not true.

3.2.2 Rankine-Hugoniot equations

Let us consider a shock front separating two regular solutions $\mathbf{u}_1(x^\beta)$ on the left side and $\mathbf{u}_o(x^\beta)$ on the right side of the shock surface and we call $\mathbf{u}_-$ and $\mathbf{u}_+$ respectively the limit of $\mathbf{u}_1$ and $\mathbf{u}_o$ on the shock surface. We can prove that the pair $(\mathbf{u}_o, \mathbf{u}_1)$ is a weak solution of (3) if and only if across the shock front the compatibility conditions (Rankine-Hugoniot equations) are verified:

$$[\![\mathbf{F}^\alpha(\mathbf{u})]\!] \, \nu_\alpha = 0 \tag{54}$$

Proof: Let σ the shock front separating the domain C in two subdomains C^+ and C^- of boundary Σ^+ and Σ^- respectively (see fig.1)

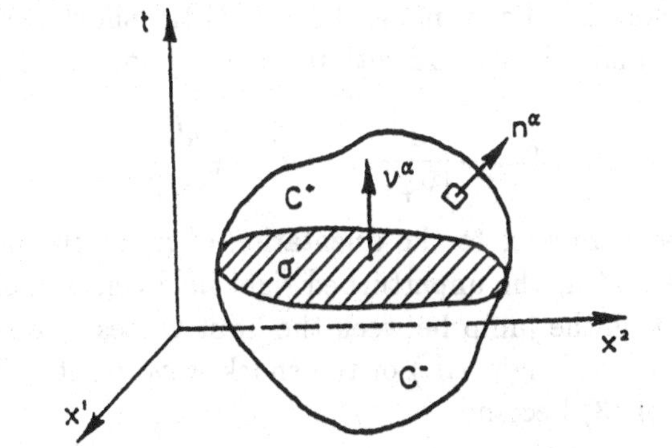

fig.1

We rewrite (52) for both the subdomains:

$$\int_{\Sigma+} n_\alpha \Phi \mathbf{F}^\alpha \, d\Sigma + \int_{\sigma+} \nu_\alpha^+ \Phi \mathbf{F}_+^\alpha \, d\sigma - \int_{C+} (\mathbf{F}^\alpha \partial_\alpha \Phi + \mathbf{f}\Phi) \, dC = 0; \tag{55}$$

$$\int_{\Sigma-} n_\alpha \Phi \mathbf{F}^\alpha \, d\Sigma + \int_{\sigma-} \nu_\alpha^- \Phi \mathbf{F}_-^\alpha \, d\sigma - \int_{C-} (\mathbf{F}^\alpha \partial_\alpha \Phi + \mathbf{f}\Phi) \, dC = 0, \tag{56}$$

where $\mathbf{F}_+^\alpha = \mathbf{F}^\alpha(\mathbf{u}_+)$, $\mathbf{F}_-^\alpha = \mathbf{F}^\alpha(\mathbf{u}_-)$. As Φ vanishes in Σ, (55) and (56) may be written

$$\int_{\sigma+} \nu_\alpha^+ \Phi \mathbf{F}_+^\alpha \, d\sigma - \int_{C+} (\mathbf{F}^\alpha \partial_\alpha \Phi + \mathbf{f}\Phi) \, dC = 0; \tag{57}$$

$$\int_{\sigma-} \nu_\alpha^- \Phi \mathbf{F}_-^\alpha \, d\sigma - \int_{C-} (\mathbf{F}^\alpha \partial_\alpha \Phi + \mathbf{f}\Phi) \, dC = 0. \tag{58}$$

Adding (58) to (57) we have:

$$\int_{\sigma+} \nu_\alpha^+ \Phi \mathbf{F}_+^\alpha \, d\sigma + \int_{\sigma-} \nu_\alpha^- \Phi \mathbf{F}_-^\alpha \, d\sigma - \int_{C} (\mathbf{F}^\alpha \partial_\alpha \Phi + \mathbf{f}\Phi) \, dC = 0, \tag{59}$$

and the last integral vanishes because we suppose that our solution is a weak solution satisfying (53). Therefore taking into account that $\nu^\alpha = \nu_+^\alpha = -\nu_-^\alpha$ we have:

$$\int_\sigma \nu^\alpha \left(\mathbf{F}_+^\alpha - \mathbf{F}_-^\alpha\right) \Phi \, d\sigma = 0. \tag{60}$$

As (60) is valid for a generic surface σ, we obtain finally:

$$\nu^\alpha \left(\mathbf{F}_+^\alpha - \mathbf{F}_-^\alpha\right) = 0, \tag{61}$$

i.e. the condition (54). Therefore the *Rankine-Hugoniot* equations constitute the compatibility conditions for the existence of shocks (particular class of weak solutions) in balance laws systems of type (3).
We introduce now more familiar simbols. Let $\varphi(x^\alpha)$ the shock surface equation and denote with s the normal velocity and with $\mathbf{n}$ the unity normal in space having then:

$$\nu^o = -\frac{s}{|\nabla\varphi|}; \qquad \nu^i = \frac{n^i}{|\nabla\varphi|}. \tag{62}$$

Moreover we call (with respect to the orientation of $\mathbf{n}$) $\mathbf{u}_l$ the perturbed left field on the shock surface and $\mathbf{u}_r$ the unperturbed right field (instead of $\mathbf{u}_-$ and $\mathbf{u}_+$) and so that $[\![\beta]\!] = \beta_l - \beta_r$ is the jump between the limit values of a generic quantity β on the left (β_l) and on the right (β_r) of the shock wave front. Then the Rankine-Hugoniot equations of (3) become

$$-s\,[\![\mathbf{F}^o]\!] + [\![\mathbf{F}^i]\!]\, n_i = 0, \tag{63}$$

i.e.

$$-sF^{o}(u_l)+F^{i}(u_l)n_i=-sF^{o}(u_r)+F^{i}(u_r)n_i. \tag{64}$$

Introducing the following mapping

$$\Psi_s(u)=-sF^{o}(u)+F_n(u)\,,\qquad F_n=F^{i}n_i, \tag{65}$$

the condition (64) for the shocks can be written as

$$\Psi_s(u_l)=\Psi_s(u_r) \tag{66}$$

which implies that, in order to have a non trivial shock ($u_l \neq u_r$), the application Ψ_s must be non locally invertible. The Jacobian of the transformation (65)

$$\frac{\partial \Psi_s}{\partial u}=-sA^{o}+A_n, \tag{67}$$

becomes singular in the point u (for a fixed unit normal n), when the shock velocity s approaches a characteristic velocity $\lambda(u)$ (see (4)):

$$s=\lambda(u). \tag{68}$$

Therefore non trivial shocks may be seen as the bifurcating branches from the trivial solution $u_l=u_r$ as sketched in fig. 2 ($\lambda_r=\lambda(u_r)$).

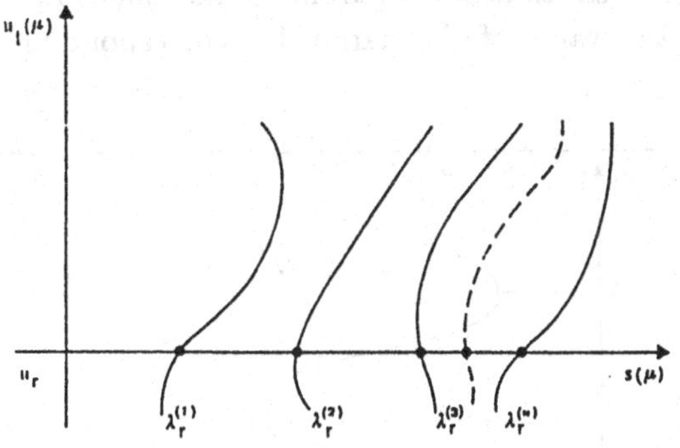

fig.2

For plane shocks (n = const.), the Rankine-Hugoniot conditions are a system of N equations for the $N+1$ unknowns u_l (*perturbed field*) and s in terms of the assigned *unperturbed field* u_r. So the usual shocks are one-parameter families of the type

$$u_l\equiv u_l(u_r,\mu)\,,\qquad s\equiv s(u_r,\mu) \tag{69}$$

where μ is the parameter.

In addition to the usual shocks there exist sometimes special shocks called *characteristic shocks.* In this circumstance we have that the characteristic velocity is continuous across the shock and coincides with the shock velocity, i.e.:

$$s = \lambda(\mathbf{u}_l) = \lambda(\mathbf{u}_r); \qquad \mathbf{u}_l \neq \mathbf{u}_r. \tag{70}$$

This corresponds to a vertical bifurcation in fig. 2. In this case it is possible to prove [2] that the shock depends on m parameters where m is the multiplicity of the eigenvalue λ:

$$\mathbf{u}_l \equiv \mathbf{u}_l(\mathbf{u}_r, \mu_I); \qquad I = 1, 2, \ldots, m. \tag{71}$$

There exists the following theorem due to G. Boillat [2]:

Theorem 3.1 *Necessary condition for the existence of a characteristic shock is that the corresponding eigenvalue λ satisfies the exceptionality condition* (9).

Of course in the linear case all the shocks are exceptional. A typical example in the non linear case is the *contact shock* in fluid dynamics for which the shock has velocity $s = \mathbf{v} \cdot \mathbf{n}$.

Now we concentrate our attention to normal shocks not satisfying the exceptionality condition and we discuss the selection rules for identifying the physical shocks.

3.2.3 Selection rules for physical shocks

Let us consider a generic shock departing from the null shock in the point A ($s(\mu_o) = \lambda_r$) of fig.3. (μ_o is the value of the parameter corresponding to the null shock: $\mathbf{u}_l(\mathbf{u}_r, \mu_o) = \mathbf{u}_r$).

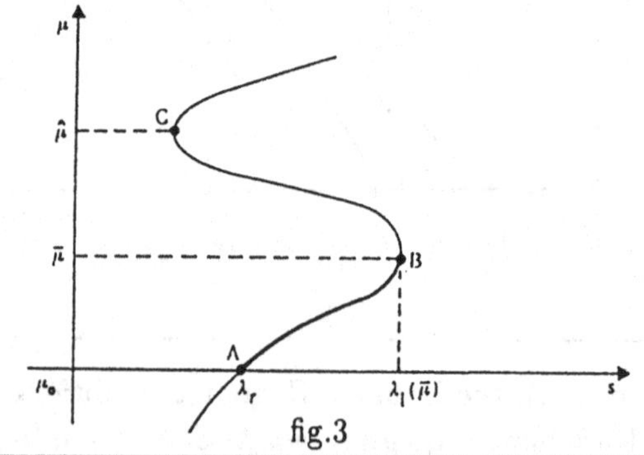

fig.3

For some value $\overline{\mu}$ of the parameter it may happen that the shock velocity is equal to the characteristic velocity evaluated in the perturbed field, i.e. $\lambda_l = \lambda(\mathbf{u}_l)$:

$$s(\overline{\mu}) = \lambda_l(\overline{\mu}). \tag{72}$$

Also in this case the Jacobian of $\mathbf{\Psi}_s$ becomes singular and then another bifurcation point appears (point B in fig.3): as a consequence, for $\mu > \overline{\mu}$ a new non trivial shock exists. We can see that in the case of fig.3 the point B corresponds to a maximum of s as function of μ (in the following the dot indicates the derivative with respect to μ, for a fixed unperturbed field): $\dot{s}(\overline{\mu}) = 0$; $\ddot{s}(\overline{\mu}) < 0$. In fig.3 we have, for $\mu = \widehat{\mu}$, another possible bifurcation point C corresponding to a minimum for s: $\dot{s}(\widehat{\mu}) = 0$; $\ddot{s}(\widehat{\mu}) > 0$.

3.2.4 Lax conditions

To eliminate the possibility of having two (or more) non trivial shocks, we can use of a first selection rule for physical shocks, the so called *Lax conditions.* They impose that the admissible shocks are those passing through the null shock and for which the shock velocity is greater than the unperturbed characteristic velocity and less than the perturbed one, [16](2) i.e.:

$$\lambda_r < s < \lambda_l; \qquad \lim_{s\to\lambda_r} \mu = \mu_o. \tag{73}$$

Therefore in the case of fig.3 the physical shocks are those lying on AB and the possible values of the parameter satisfy(3):

$$\mu_o < \mu < \overline{\mu}. \tag{74}$$

We call these shocks *positive shocks.*

The situation described in fig.3 depends on the choice of the unperturbed field $\mathbf{u}_r$. In fact for a different value of $\mathbf{u}_r$ we may have a similar situation as shown in fig.4: now the Lax conditions select as possible shocks those lying on AB (*negative shocks*) such that:

$$\overline{\mu} < \mu < \mu_o \tag{75}$$

(2)In terms of unperturbed and perturbed Mach numbers ($M_r = s/\lambda_r$ and $M_l = s/\lambda_l$, respectively) the Lax conditions are equivalent to the well known conditions, arising in fluid dynamics, that the shock is supersonic at one side ($M_r > 1$), and subsonic at the other one ($M_l < 1$).

(3)Of course it is possible to have cases in which $\overline{\mu} \to \infty$. In this circumstance we don't have other bifurcations points except the point A (e.g. the fluid dynamics).

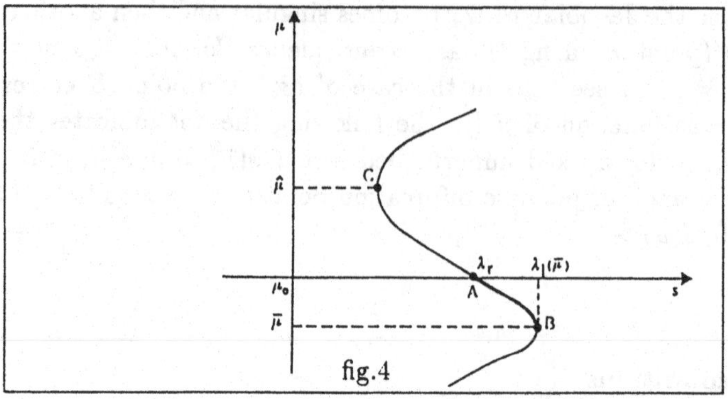

fig.4

Moreover values of the unperturbed field $\mathbf{u}_r$ can exist for which two very special cases, as in figg.5 and 6, occur:

- a) in fig.5 the null shock A coincides with the point B ($\mu_o = \overline{\mu}$) and so no shocks are admissible;

- b) in fig.6 the null shock A coincides with the minimum point C ($\mu_o = \widehat{\mu}$) and both the positive and negative shocks are possible.

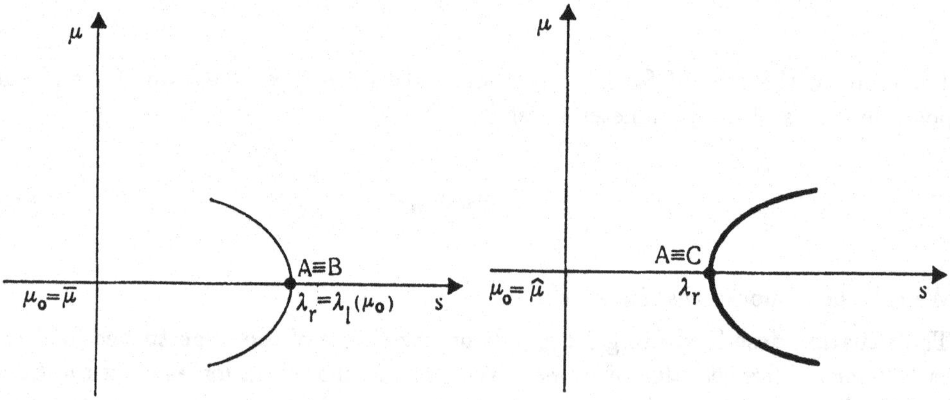

fig.5 fig.6

These special cases are very important for our analysis (in particular the case a)) and for this reason it is interesting to find if there exist possible values of the unperturbed field $\mathbf{u}_r$ for which these extreme cases can be verified. The answer to this question lies in the following theorem [17],

Theorem 3.2 *The special cases a), b) are possible only if there exists* $\tilde{\mathbf{u}}_r$*, such that:*

$$(\nabla\lambda \cdot \mathbf{d})_{\tilde{\mathbf{u}}_r} = 0, \tag{76}$$

where $\nabla = \partial/\partial \mathbf{u}$, *and* λ, $\mathbf{d}$ *are respectively an eigenvalue and the corresponding right eigenvector of the characteristic eigenvalue problem* (4):

$$(\mathbf{A}_n - \lambda \mathbf{A}^o)\,\mathbf{d} = 0. \tag{77}$$

In particular, (76) *is equivalent (in the shock) to the following condition*

$$\dot{\lambda}_{l_o} = \lim_{\mu \to \mu_o} \frac{d\lambda_l}{d\mu}(\tilde{\mathbf{u}}_r, \mu) \propto (\nabla\lambda \cdot \mathbf{d})_{\tilde{\mathbf{u}}_r} = 0 \tag{78}$$

and the case a) (no admissible shocks) occurs if λ_l *has a maximum in* μ_o *while the case b) (two shocks families admissible) occurs if* λ_l *has a minimum in* μ_o.

The condition (76) is very special because in the usual examples (e.g. the fluid dynamics case) we have either waves satisfying the *genuine non-linearity* (8) or eigenvalues satisfying the *exceptionality condition* (9) The condition (76) violates both the previous conditions, because now the equality to zero occurs only for a particular field value. We finish about these selection rules with the following observations:

1. we have written before the Lax conditions for a generic eigenvalue λ but, in reality, taking into account all the eigenvalues (that we suppose distinct), the general statement can be expressed as[4]

$$\begin{aligned} &\lambda_r^{(1)} < \lambda_r^{(2)} < \ldots < \lambda_r^{(k)} < s < \lambda_r^{(k+1)} < \ldots < \lambda_r^{(N)}; \\ &\lambda_l^{(1)} < \lambda_l^{(2)} < \ldots < \lambda_l^{(k-1)} < s < \lambda_l^{(k)} < \ldots < \lambda_l^{(N)}; \\ &\lim_{s \to \lambda_r^{(k)}} [\![\mathbf{u}]\!] = 0; \qquad (k = 1, 2, \ldots, N). \end{aligned} \tag{79}$$

   implying, in particular, the condition (73):

$$\lambda_r^{(k)} < s < \lambda_l^{(k)}; \qquad \lim_{s \to \lambda_r^{(k)}} [\![\mathbf{u}]\!] = 0. \tag{80}$$

   For a fixed k the shock satisfying (79) is called a *k-shock.* When k assumes the values $1, 2, \ldots, N$, the conditions (79) select in this way N shocks in correspondence of the N eigenvalues $\lambda_r^{(k)}$.

2. As well known, weak solutions don't have uniqueness for initial data of Riemann type. If the initial jump does not satisfy the Lax conditions it is possible to conjecture that after the initial time a regular solution appears. In fact in the homogeneous hyperbolic systems (non dissipative), it is well known (see e.g. [20]) that for Riemann data violating the Lax conditions, we have a regular solution of simple wave type after $t = 0$. In the dissipative case a general proof for the regularization does not exist, but it is even more acceptable to think that the same thing happens.

(4) The Lax conditions play several roles in non-linear wave propagation; in particular for the causality in the impact from a discontinuity wave and a shock waves see e.g. [18, 19]

3.2.5 Entropy growth condition

Another criterion, instead the Lax conditions, for selecting physical shocks among all the solutions of eqs. (63) is furnished by the thermodynamical requirement of *entropy growth across the shock*. The equations (63) are formally obtained from the field eqs. (3) by the correspondence rule

$$\partial_t * \to -s\,[\![*]\!] \qquad \partial_i * \to n_i\,[\![*]\!] \tag{81}$$

and the production terms in (3) do not play any role in the Rankine-Hugoniot equations. Also if the differentiable solutions of (3) are compatible with the entropy inequality (10), the rule (81) does not hold when it is applied to (10). In fact there exists an entropy production across the shock

$$\eta = -s\,[\![h^o]\!] + [\![h^i]\!]\,n_i \tag{82}$$

generally non vanishing, in the non linear case, for the solutions (69) of the Rankine-Hugoniot equations (63). The choice

$$\eta(\mathbf{u}_r, \mu) > 0, \tag{83}$$

is often referred in literature as *entropy growth condition* [21, 22] and is assumed as a criterion to peak up physical shocks among the solutions of the Rankine-Hugoniot equations. We remark that the circumstance η non vanishing, means that while the law (10) follows from the field equations when differentiability conditions hold, it does not follow for the weak solutions (as shock waves are). It is well known that

Theorem 3.3 *For weak shocks satisfying the genuine non linearity* (8) *the Lax conditions and the entropy growth are equivalent.*

It is possible to prove that also in the special cases considered in the theorem 3.2 in which the (76) holds the Lax conditions and the entropy growth give the same results for weak shocks. In fact (see [17]):

Theorem 3.4 *In the case a) of the theorem 3.2 (no shocks for the Lax conditions), the entropy across the shock η is always non positive in a neighbourhood of the null shock and in the case b) (two shocks admissible for Lax) the function η is always non negative in a neighbourhood of the null shock. Therefore the case a) is not admissible while the case b) is always satisfied also for the entropy growth condition.*

Therefore for weak shocks we have coincidence between the two selection rules for physical shocks[(5)]. In the case of strong shocks the two criteria are, in general, not equivalent as we can see in the second sound problem in Part IV.

[(5)] A discussion of the selection rules and, more in general, of the questions related to the Riemann problem can be seen in C. Dafermos and in references therein quoted [23].

II. Acceleration and Shock Waves in Extended Thermodynamics

4 Extended Thermodynamics of an Ideal Gas

Using only universal principle in the spirit of *Rational Thermodynamics* it was possible to construct a continuum approach to non equilibrium thermodynamics of gases (Extended Thermodynamics [24]). The surprising result was that only using macroscopic arguments the so obtained field equations coincide, in the case of the ideal gases, with those deduced by means of the Grad 13-moments procedure from the Boltzmann equation. For simplicity, we write down the equations in the unidimensional case (the dot indicates the material time derivative):

$$\begin{cases} \dot{\rho} + \rho v_x = 0 \\ \rho \dot{v} + (p - \sigma)_x = 0 \\ \rho \dot{e} + q_x + (p - \sigma) v_x = 0 \\ \tau_\sigma \left[\dot{\sigma} - \frac{8}{15} q_x + \frac{7}{3} \sigma v_x \right] - \frac{4}{3} \mu v_x = -\sigma \\ \tau_q \left[\dot{q} + \frac{16}{5} q v_x - \frac{7}{2} (\frac{p}{\rho})_x \sigma - \frac{1}{\rho} (p + \sigma) \sigma_x + \frac{\sigma}{\rho} p_x \right] + \chi \theta_x = -q \end{cases} \tag{84}$$

$$\tau_\sigma = \frac{\mu}{p}; \quad \tau_q = \frac{2}{5} \frac{\chi}{p^2} \rho \theta; \quad \frac{\tau_\sigma}{\tau_q} = \frac{2}{3}; \quad p = \Re \rho \theta. \tag{85}$$

The first three equations are the usual conservation laws while the other new equations reduce to the Fourier and Navier Stokes equations when the relaxing times τ_σ and τ_q vanish. The simbols are the usual ones: $\rho, v, p, e, q, \sigma, \mu$ and χ are respectively the mass density, velocity, pressure, internal energy, heat flux, shear stress, viscosity coefficient and heat conductivity.

The coincidence in the case of ideal gases between our approach and the microscopic one gives a good test on the validity of the extended thermodynamics arguments with the advantage that it is possible to use this approach also for non ideal gases, solids, etc...

5 Thermo - Acceleration Waves

In this section we apply the general results of the previous chapter sect. 3.1 to study the temporal evolution of the discontinuity waves associated to the system (84). At first, we evaluate the characteristic velocities of (84) in a generic non equilibrium state.

5.1 Characteristic velocities

As it was stated in the previous chapter the characteristic velocities for the general system

$$\mathbf{u}_t + \mathbf{A}(\mathbf{u})\mathbf{u}_x = \mathbf{f}(\mathbf{u}) \tag{86}$$

are solutions of the eigenvalue problem

$$(\mathbf{A} - \lambda \mathbf{I})\mathbf{d} = 0. \tag{87}$$

To obtain (87) from (86) it is possible to apply the correspondence rule

$$\partial_t \to -\lambda\delta; \quad \partial_x \to \delta; \quad \mathbf{f} \to 0, \tag{88}$$

where δ is a differential operator. In this way, we get (87) from (86) with

$$\delta\mathbf{u} \propto \mathbf{d}. \tag{89}$$

Since the system (84) is written by using the material time derivative, the corresponding rule is

$$\cdot = \partial_t + v\partial_x \to -\overline{\lambda}\delta, \tag{90}$$

where

$$\overline{\lambda} = \lambda - v \tag{91}$$

is the relative velocity of the wave with respect to the fluid velocity.
Therefore from (84), (88) and (90) it follows

$$\begin{cases} -\overline{\lambda}\delta\rho + \rho\delta v = 0 \\ -\rho\overline{\lambda}\delta v + \delta(p - \sigma) = 0 \\ -\rho\overline{\lambda}\delta e + \delta q + (p - \sigma)\delta v = 0 \\ -\overline{\lambda}\delta\sigma - \frac{8}{15}\delta q + \frac{7}{3}\sigma\delta v - \frac{4}{3}p\delta v = 0 \\ -\overline{\lambda}\delta q + \frac{16}{5}q\delta v - \frac{7}{2}\sigma\delta\left(\frac{p}{\rho}\right) - \frac{1}{\rho}(p + \sigma)\delta\sigma + \frac{\sigma}{\rho}\delta p + \frac{5}{2}\frac{p^2}{\rho\theta}\delta\theta = 0. \end{cases} \tag{92}$$

Taking into account the constitutive relations (85) and introducing the sound velocity c

$$c = \sqrt{\frac{5}{3}\mathcal{R}\theta}, \tag{93}$$

it is possible to put the previous system in the form of the linear homogeneous system (87):

$$\left(\overline{\mathbf{A}} - \overline{\lambda}\mathbf{I}\right)\delta\mathbf{u} = 0, \tag{94}$$

with

$$\mathbf{u} \equiv (\rho, v, \theta, \sigma, q)^T, \tag{95}$$

and

$$\overline{\mathbf{A}} \equiv \begin{pmatrix} 0 & \rho & 0 & 0 & 0 \\ \frac{3c^2}{5\rho} & 0 & \frac{3c^2}{5\theta} & -\frac{1}{\rho} & 0 \\ 0 & \frac{2}{9}\theta(3-5\widehat{\sigma}) & 0 & 0 & \frac{10}{9}\frac{\theta}{c^2\rho} \\ 0 & c^2\rho\left(\frac{7}{3}\widehat{\sigma} - \frac{4}{5}\right) & 0 & 0 & -\frac{8}{15} \\ \frac{3}{5}c^4\widehat{\sigma} & \frac{16}{5}c^3\rho\widehat{q} & -\frac{3}{10}\frac{c^4\rho}{\theta}(5\widehat{\sigma}-3) & -\frac{c^2}{5}(5\widehat{\sigma}+3) & 0 \end{pmatrix}. \tag{96}$$

Here we have introduced the dimensionless parameters $\widehat{\sigma}$ and $\widehat{q}$ related to the shear stress and the heat flux through:

$$\widehat{\sigma} = \frac{\sigma}{c^2\rho}; \qquad \widehat{q} = \frac{q}{c^3\rho}. \tag{97}$$

By requiring

$$\det\left(\overline{\mathbf{A}} - \overline{\lambda}\mathbf{I}\right) = 0, \tag{98}$$

the following characteristic polynomial yields:

$$\widehat{\lambda}\left\{\widehat{\lambda}^4 + \widehat{\lambda}^2\left(\frac{62}{15}\widehat{\sigma} - \frac{78}{25}\right) - \frac{96}{25}\widehat{q}\widehat{\lambda} + \frac{21}{5}\widehat{\sigma}^2 - \frac{18}{5}\widehat{\sigma} + \frac{27}{25}\right\} = 0, \tag{99}$$

where

$$\widehat{\lambda} = \frac{\overline{\lambda}}{c}. \tag{100}$$

From (99) we obtain the *contact wave*

$$\widehat{\lambda} = 0 \quad \text{i.e.} \quad \lambda = v, \tag{101}$$

and four *thermo-viscous-sonic waves* solutions of

$$\widehat{\lambda}^4 + \widehat{\lambda}^2\left(\frac{62}{15}\widehat{\sigma} - \frac{78}{25}\right) - \frac{96}{25}\widehat{q}\widehat{\lambda} + \frac{21}{5}\widehat{\sigma}^2 - \frac{18}{5}\widehat{\sigma} + \frac{27}{25} = 0. \tag{102}$$

From (102) we can see that $\widehat{\lambda}$ depends on the choice of the non equilibrium parameters $\widehat{\sigma}$ and $\widehat{q}$. Therefore the discontinuity waves propagating in a generic non equilibrium state have velocities that are the roots of (102). In the equilibrium state, (102) reduces to the biquadratic equation

$$\widehat{\lambda}_o^4 - \frac{78}{25}\widehat{\lambda}_o^2 + \frac{27}{25} = 0 \tag{103}$$

and according with the general results of the previous chapter these velocities coincide with the phase velocities in the high frequency limit:

$$\widehat{\lambda}_o \simeq \pm 1.6503, \quad \widehat{\lambda}_o \simeq \pm 0.6297. \tag{104}$$

We observe that at non equilibrium the waves travelling in the $x > 0$ region have velocities different with respect to the waves going in the opposite direction. This fact is due to the presence of heat flux $\widehat{q}$ appearing in the linear coefficient of (102). Moreover we recall that our theory is valid only for processes not far from equilibrium and therefore is not surprising that the system is hyperbolic only in a neighbourhood of the equilibrium state i.e. when $\widehat{q}$ and $\widehat{\sigma}$ are sufficiently small. For what concerns the hyperbolicity, an analysis of the characteristic polynomial (102) permits to determine the radius of the circular neighbourhood of the equilibrium state in which the roots of (102) are all real and therefore we have a *measure* of the validity of the theory.

5.2 Hyperbolicity region

We want to determine the region in the plane $(\widehat{q}, \widehat{\sigma})$ in which the system is hyperbolic, i.e. all the roots of (102) are real. To this aim, we observe that the boundary separating the hyperbolicity region from the region with complex roots is necessarily the set of points such that two roots become coincident. In fact if two adjacent roots are at first real and then become complex it is necessary, for continuity reasons, that there exists a point in which the two roots coincide. Therefore to find this boundary we require, in the plane $(\widehat{q}, \widehat{\sigma})$, the existence of points such that the polynomial (102) is of the form:

$$(\widehat{\lambda} - \mu_1)^2(\widehat{\lambda} - \mu_2)(\widehat{\lambda} - \mu_3) = 0. \tag{105}$$

Introducing the parameters

$$a = \frac{1}{2}(\mu_2 + \mu_3); \qquad b = \frac{1}{2}(\mu_2 - \mu_3), \tag{106}$$

the equivalence between (102) and (105) implies $\mu_1 = -a$ and

$$\frac{62}{15}\widehat{\sigma} - \frac{78}{25} = -2a^2 - b^2; \tag{107}$$

$$\frac{21}{5}\widehat{\sigma}^2 - \frac{18}{5}\widehat{\sigma} + \frac{27}{25} = a^2(a^2 - b^2); \tag{108}$$

$$\frac{96}{25}\widehat{q} = 2ab^2. \tag{109}$$

From (107) and (108) we obtain with some algebra:

$$a^2 = \frac{1}{225}\left(\sqrt{2}\Delta - 155\widehat{\sigma} + 117\right); \quad b^2 = -\frac{2\sqrt{2}}{225}\left[\Delta + \sqrt{2}(155\widehat{\sigma} - 117)\right], \tag{110}$$

where we put

$$\Delta = \sqrt{47450\widehat{\sigma}^2 - 48510\widehat{\sigma} + 15957}.$$

From (109):

$$\widehat{q} = \frac{25}{48}ab^2, \tag{111}$$

and inserting a e b from (110) in this expression we obtain $\widehat{q}$ as function of $\widehat{\sigma}$. Observe that if μ_2 and μ_3 are real it follows that a and b are real and so the second members of (110) must be positive this implying that $\widehat{\sigma} < \widehat{\sigma}^* \simeq .481060$. If μ_2 and μ_3 are complex conjugate a is real but b is pure imaginary and the second member of $(110)_2$ is negative, i.e. $\widehat{\sigma} > \widehat{\sigma}*$. Using these considerations we plot the results in fig. 7, where it is evident the existence of four regions:

- the hyperbolicity region, in which the roots are all real and distinct (strictly hyperbolicity), upper bounded from $\widehat{\sigma}^*$ and lower unbounded ;

- the region II in which two roots are real and two complex. In particular in the right side of fig.7 the two real roots are the positive ones, while in the left side the real are the negative ones. On the boundary from the hyperbolicity region to the region II two roots are real and coincident. Therefore this boundary represents the points in which we have non strictly hyperbolicity. In the shock problems these points are called *umbilic points.*

- the region IV in which all roots are complex.

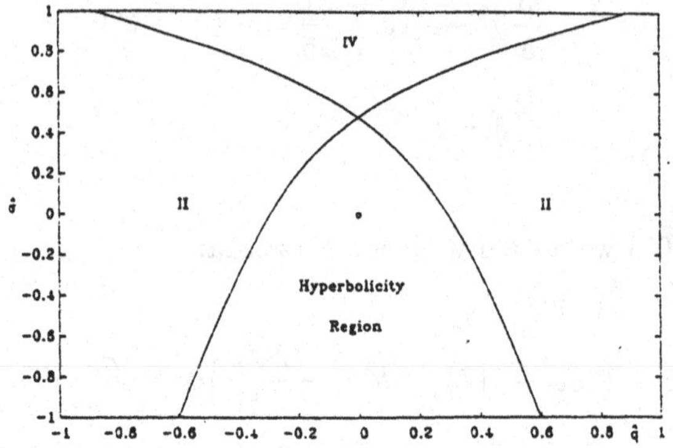

fig.7

Now it is a simple matter to construct the equation of maximum circle centred in the origin and contained in the hyperbolicity region. This is done in fig. 8 where the internal region to this circle represents the maximum circular neighbourhood in which the Extended Thermodynamics theory is valid for what concerns the hyperbolicity. The hyperbolicity radius evaluated is

$$r_{\rm hyp}\sqrt{\widehat{q}^2+\widehat{\sigma}^2}=\frac{1}{\hat{c}^2\rho}\sqrt{\frac{q^2}{c^2}+\sigma^2}\simeq 0.2686. \tag{112}$$

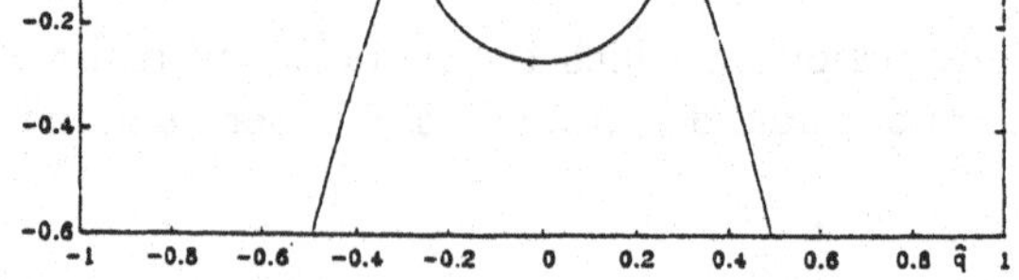

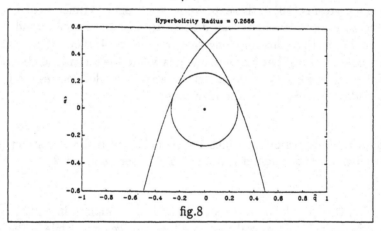

fig.8

5.3 The evolution law

Let's now evaluate the temporal behaviour of a discontinuity wave amplitude following the results stated in sect.3.1 of Part I.
We choose as unperturbed state, in which the wave propagates, the equilibrium constant state:

$$\mathbf{u}_o \equiv (\rho_o, v_o, \theta_o, \sigma_o, q_o)^T; \qquad v_o = \sigma_o = q_o = 0, \tag{113}$$

and we consider the fastest wave propagating with velocity $\lambda_o = 1.6503c_o$. First of all we find the right and the left eigenvectors of the matrix (96) evaluated in the equilibrium state. Simple calculations give:

$$\mathbf{d}_o \equiv \left(1, c\frac{\widehat{\lambda}}{\rho}, \frac{\theta}{27\rho}\left(25\widehat{\lambda}^2 - 27\right), -\frac{4c^2\widehat{\lambda}^2}{9}, \frac{c^3\widehat{\lambda}}{6}\left(5\widehat{\lambda}^2 - 9\right)\right)_o^T; \tag{114}$$

$$\mathbf{l}_o \equiv \alpha\left(\frac{1}{25}c^3\left(25\widehat{\lambda}^2 - 33\right), \frac{1}{15}c^2\widehat{\lambda}\rho\left(25\widehat{\lambda}^2 - 33\right), \frac{1}{25\theta}c^3\rho\left(25\widehat{\lambda}^2 + 12\right), -\frac{1}{3}c\left(5\widehat{\lambda}^2 - 3\right), 2\widehat{\lambda}\right)_o \tag{115}$$

with

$$\alpha = \left(\frac{5}{4c^3\left(13\widehat{\lambda}^2 - 9\right)}\right)_o, \tag{116}$$

such that $\mathbf{l} \cdot \mathbf{d} = 1$.
With the previous choice of the right eigenvector we have (see (89)):

$$\delta\mathbf{u} = \delta\rho\ \mathbf{d}_o; \qquad \delta = \left[\left[\frac{\partial}{\partial\varphi}\right]\right], \tag{117}$$

and the coefficients a and b (3.5),(3.6) of Bernoulli transport equation (3.4) become in the constant state:

$$a = (\nabla\lambda \cdot \mathbf{d})_o = \frac{1}{\delta\rho}\nabla\lambda \cdot \delta\mathbf{u} = \left(\frac{\delta\lambda}{\delta\rho}\right)_o; \tag{118}$$

$$b = -(\mathbf{l} \cdot \nabla\mathbf{f} \cdot \mathbf{d})_o = -\frac{1}{\delta\rho}(\mathbf{l} \cdot \nabla\mathbf{f} \cdot \delta\mathbf{u})_o = -\left(\frac{1}{\delta\rho}\mathbf{l} \cdot \delta\mathbf{f}\right)_o \tag{119}$$

Differentiating (102) and taking into account (117), (114),(100) and (91), we obtain with some algebra:

$$a = \frac{c_o}{\rho_o}\alpha_o; \qquad \alpha_o = \left(\widehat{\lambda}\frac{625\widehat{\lambda}^4 + 1400\widehat{\lambda}^2 - 3537}{54\left(25\widehat{\lambda}^2 - 39\right)}\right)_o \simeq 5.16061. \tag{120}$$

Taking into account that the production vector of our system is

$$\mathbf{f} \equiv \left(0, 0, 0, -\frac{\sigma}{\tau_\sigma}, -\frac{q}{\tau_q}\right); \qquad \tau_\sigma = \frac{2}{3}\tau_q, \tag{121}$$

we get from (119):

$$b = \frac{\beta_o}{\tau_q^o}; \qquad \beta_o = \frac{5}{4}\left(\frac{5\widehat{\lambda}^3 - 3}{13\widehat{\lambda}^2 - 9}\right) \simeq 0.5027, \tag{122}$$

that coincide, according with the general theory of the previous chapter, with the temporal absorption of the linear plane wave [8]. From sect. 3.1 and (117) one has:

$$\Pi = \left[\!\!\left[\frac{\partial \mathbf{u}}{\partial \varphi}\right]\!\!\right] = \delta \mathbf{u} = \delta\rho\ \mathbf{d}_o. \tag{123}$$

Therefore the amplitude Π of the Bernoulli equation coincides in our case with the normal derivative jump of the density:

$$\Pi = \delta\rho. \tag{124}$$

It is convenient to rewrite it in terms of the acceleration jump $G = [\![v_t]\!]$. Taking into account (114), the following identities arise:

$$\begin{aligned} G &= [\![v_t]\!] = [\![v_\varphi]\!]\,\varphi_t = -[\![v_\varphi]\!]\,\lambda_o\varphi_x = \\ &= -\hat{\lambda}c_o\delta v = -\hat{\lambda}_o^2 c_o^2\frac{\delta\rho}{\rho_o} = -\hat{\lambda}_o^2 c_o^2\frac{\Pi}{\rho_o}. \end{aligned} \tag{125}$$

Therefore the solution (3.8) gives the temporal evolution law of the acceleration jump:

$$G(t) = \frac{G(0)e^{-\beta_o\tau}}{1 - \frac{G(0)\tau_q^o}{c_o}\frac{\alpha_o}{\beta_o\hat{\lambda}_o^2}\left(1 - e^{-\beta_o\tau}\right)}; \qquad \tau = \frac{t}{\tau_q^o}. \tag{126}$$

Introducing the dimensionless acceleration:

$$\Gamma = G\frac{\tau_q^o}{c_o}, \tag{127}$$

and putting

$$\mu = \frac{\alpha_o}{\beta_o\hat{\lambda}_o^2} \simeq 6.22164, \tag{128}$$

the law (126) may be written as:

$$\Gamma = \frac{\Gamma(0)e^{-\beta_o\tau}}{1 - \Gamma(0)\mu\left(1 - e^{-\beta_o\tau}\right)}. \tag{129}$$

From (129), if

$$\Gamma(0) > \Gamma^* = \frac{1}{\mu} \simeq 0.160729, \tag{130}$$

then there exists a critical time

$$\tau_{\text{crit}} = -\frac{1}{\beta_o}\log\left(1 - \frac{1}{\Gamma(0)\mu}\right), \tag{131}$$

such that

$$\lim_{\tau \to \tau_{crit}} \Gamma(\tau) = \infty, \tag{132}$$

and a shock takes place. On the contrary if

$$\Gamma(0) < \Gamma^* \tag{133}$$

the amplitude $\Gamma(t)$ decays to zero when the time increases. This means that, in order to have a shock formation, it is necessary that $\Gamma(0) > 0$ (compression wave) and this amplitude must be sufficiently large $\Gamma(0) > \Gamma^*$. Vice versa for weak initial amplitude one has $\Gamma(0) < \Gamma^*$, $\lim_{\tau \to \infty} \Gamma(\tau) = 0$ according with the general results of sect. 3.1 stating that if b is positive we have linear and non linear λ–stability (asymptotic stability). We observe that the condition for the existence of a shock formation (130) implies:

$$G(0) > \frac{c_o}{\tau_q^o} \Gamma^* \tag{134}$$

and so taking into account that the relaxing time τ_q^o is very small it is necessary to have an enormous initial acceleration to have a critical time. Therefore in normal life any initial acceleration wave is regularised in a very brief time. Of course in some astrophysical scale it should be possible, at least in principle, to have a shock formation as degeneration of an acceleration wave. The interested reader may read the paper [25, 26]. In particular in [26] the most general case of acceleration waves for a gas in a gravitational field is considered.

6 Shocks in Extended Thermodynamics

The Extended Thermodynamics is a physical example of an hyperbolic dissipative system of balance laws. In this framework it is possible to define weak solutions and, in particular, to study shock waves just like in a non dissipative system.
In this section we study shock waves propagating through an equilibrium state. The problem of *shock structure* (shocks with thickness) will be considered apart in the next chapter. A qualitative analysis of the Rankine-Hugoniot equations with the selection rules for admissible shocks (Lax conditions, entropy growth) gives different results with respect to the classical approach. Infact in Extended Thermodynamics it is possible to prove that just behind the shock front a non equilibrium state always exists and two fronts are present: a *hot shock*, in which the temperature behind the shock is greater than the unperturbed temperature and a *cold shock* in which the perturbed temperature is smaller than the unperturbed one. In these cases there exists a small set of Mach numbers less than one for which the admissibility criteria are satisfied. The consequences of these results are discussed, in particular pointing out that the non equilibrium temperature loses the usual role since it is no longer

a *measure* of the disorder as in the equilibrium case. Moreover we find the limit values of the Mach numbers corresponding to the so called *umbilic points*. After these values the differential system is not hyperbolic and so the theory cannot be longer applied.

6.1 Balance Laws and Rankine-Hugoniot Equations

We rewrite the equations of the Extended Thermodynamics in one space dimension in the form of a balance laws system [24]:

$$\frac{\partial}{\partial t}\rho + \frac{\partial}{\partial x}(\rho v) = 0 \tag{135}$$

$$\frac{\partial}{\partial t}(\rho v) + \frac{\partial}{\partial x}(\rho v^2 + p - \sigma) = 0 \tag{136}$$

$$\frac{\partial}{\partial t}(\rho v^2 + 3p) + \frac{\partial}{\partial x}(\rho v^3 + 5pv - 2\sigma v + 2q) = 0 \tag{137}$$

$$\frac{\partial}{\partial t}(\frac{2}{3}\rho v^2 - \sigma) + \frac{\partial}{\partial x}\left(\frac{2}{3}\rho v^3 + \frac{4}{3}pv - \frac{7}{3}\sigma v + \frac{8}{15}q\right) = \tau_o \sigma \tag{138}$$

$$\frac{\partial}{\partial t}\left(2q + 5pv - 2\sigma v + \rho v^3\right) + \frac{\partial}{\partial x}\left(\rho v^4 + 5\frac{p^2}{\rho} - 7\frac{\sigma p}{\rho} + \right.$$

$$\left. + \frac{32}{5}qv + v^2(8p - 5\sigma)\right) = 2\tau_o v\sigma - \tau_1 q, \tag{139}$$

Consider now a plane shock front with velocity s. Introducing the relative velocity

$$u = v - s, \tag{140}$$

the Rankine-Hugoniot (R. H.) compatibility conditions across the shock front become

$$[\![\rho u]\!] = 0 \tag{141}$$

$$[\![\rho u^2 + p - \sigma]\!] = 0 \tag{142}$$

$$[\![2q + 5pu - 2\sigma u + \rho u^3]\!] = 0 \tag{143}$$

$$\left[\!\!\left[\frac{8}{15}q + \frac{4}{3}pu - \frac{7}{3}\sigma u + \frac{2}{3}\rho u^3\right]\!\!\right] = 0 \tag{144}$$

$$\left[\!\!\left[\rho u^4 + 5\frac{p^2}{\rho} - 7\frac{p\sigma}{\rho} + \frac{32}{5}uq + u^2(8p - 5\sigma)\right]\!\!\right] = 0. \tag{145}$$

where $[\![f]\!] = f_1 - f_o$ indicates the jump and f_o, f_1 are the limit values of f across the shock front evaluated respectively in the unperturbed and perturbed state. We consider the propagation in an equilibrium state, i.e. when the unperturbed state is characterized by $\rho_o, p_o, v_o = q_o = \sigma_o = 0$. Introducing the dimensionless variables:

$$M_o = \frac{s - v_o}{c_o} = \frac{s}{c_o}; \; w = \frac{u_1}{u_o}; \; \pi = \frac{p_1}{p_o}; \; r = \frac{\rho_1}{\rho_o}; \; \chi = \frac{q_1}{p_o c_o}; \; \tau = \frac{\sigma_1}{p_o}, \tag{146}$$

it is possible, after straightforward calculations, to give the solution of the R. H. equations (for non characteristic shocks) in the form [27]:

$$r = \frac{1}{w} \tag{147}$$

$$\tau = \frac{10}{27} M_o^2 \frac{w^2 - 1}{w} \tag{148}$$

$$\pi = 1 - \frac{5}{27} M_o^2 \frac{(w-1)(7w-2)}{w} \tag{149}$$

$$\chi = -\frac{5}{18} M_o (w-1) \left(10 M_o^2 w - 5 M_o^2 - 9\right) \tag{150}$$

where w and M_o are related by the following polynomial relation:

$$5 M_o^4 \left(694 w^3 - 710 w^2 + 143 w + 8\right) - 2106 M_o^2 w (2w - 1) + 729 w = 0. \tag{151}$$

Therefore for a fixed value M_o of the unperturbed Mach number (characterizing the shock parameter), we obtain w from (151) and then r, τ, π and χ from (147)-(150). From (147) and (146) we note that w is the volume ratio, π is the pressure ratio, τ and χ are the dimensionless shear stress and heat flux in the non equilibrium perturbed region. Moreover taking into account (140), (146) and the state function $p = \mathcal{R}\rho\theta$, it is easy to evaluate also the velocity and the temperature just behind the shock front:

$$\overline{v}_1 = \frac{v_1}{c_o} = M_o(1 - w) \qquad \Theta = \frac{\theta_1}{\theta_o} = \pi w. \tag{152}$$

In accordance with the general theory of shock waves the shocks are one parameter families corresponding to the bifurcated branches of the trivial solution (null shock), occurring when s approaches to a characteristic eigenvalue λ of the differential system (135)-(139). In fact, observing that from (147)-(150) the null shock corresponds to $w \to 1$, (151) reduces to

$$25 M_o^4 - 78 M_o^2 + 27 = 0, \tag{153}$$

i.e. the characteristic polynomial of the differential system at equilibrium (103). The roots of (153) are

$$M_o \simeq \pm 0.6297; \quad \pm 1.65028. \tag{154}$$

Therefore considering the shocks propagating in the $x > 0$ region, we have two families of shocks bifurcating from the null solution, when M_o approaches respectively to ${M_o}^{(S)} = 0.6297$ (slow shock) and ${M_o}^{(F)} = 1.65028$ (fast shock). The following general properties are easily verified:

Statement 6.1 *Just behind the shock front we have always a non equilibrium state.*

In fact from (148), (150) the only possibility for which the heat flux and the shear stress are both zero is $w = 1$, i.e. the null shock. Another general result is the following one:

Statement 6.2 *For both the non characteristic shocks there exists a lower bound for the volume ratio w:*

$$w = \frac{V_1}{V_o} > w_{\text{crit}} \simeq 0.709949 \tag{155}$$

instead of $w_{\text{crit}} = 0.25$ of the classic case without dissipation (see e.g. [28]).

The proof of this statement follows by an analysis of the coefficients of (151) such that the biquadratic equation (151) may admit real roots M_o for a fixed value of w. When $w \to w_{\text{crit}}$ in the fast shock, M_o approaches to $+\infty$. Therefore in E.T. the volume ratio is always greater than the classical expectation.
We calculate the two positive solutions $M_o = M_o(w)$ of (151) corresponding to the slow and fast shocks. These solutions are introduced into the equations and thus provide $w, \Theta, \tau, \chi, \pi, \overline{v}_1$ as functions of s, or M_o. These functions have been numerically calculated and they are represented in [24]. All curves for the slow shock have been broken off at $M_o = 0.908$; this is the value where w reaches the value 0.709949, i.e. where the fast shock creates the maximum jump in density. Not all parts of these plots are physically relevant, because we have selections rules which imply that some parts are unphysical, although they are perfectly valid solutions of the Rankine Hugoniot equations. The Lax conditions (3.46) for the fastest wave $k = 4$ are satisfied provided $M_o > 1.6503$ holds and we have after the passage of fast shock that (see [24]):

the density jumps up

the temperature jumps up

the stress jumps down

the heat flux jumps up, and

the velocity jumps up.

The slow shock can only propagate in the range of Mach numbers $0.6297 < M_o \leq 0.908$ and we see that in this case:

the density jumps up

the temperature jumps down

the stress jumps down

the heat flux jumps down, and

the velocity jumps up.

The main interesting result of this statement is the unusual condition that the temperature after the shock is less than the unperturbed one.
The non usual *cold shock* might appear inconsistent with thermodynamics because behind the shock front we have a region of non-equilibrium, i.e. more ***disorderly***, and therefore it is reasonable to think that the temperature increases after the passage of the shock. But, as well known, the real measure of disorder is the entropy growth across the shock front and as the density of entropy at non equilibrium depends not only on the temperature but also on the heat flux and the shear stress, it is possible that the entropy increases after the passage of the shock though the temperature decreases. Therefore the cold shock is compatible with the thermodynamics principles and the temperature cannot be considered as a measure of the disorder in the non equilibrium state. This situation is also present in the phonon real gas (see Part IV).
This problem present also the interesting feature that when $M_o \approx 2.7$ the system looses hyperbolicity after the passage of a fast shock, because two eigenvalues become equal in the disturbed region. We have discussed this break-down of hyperbolicity in sect. 5.2.

III. The problem of Shock Wave Structure

In the previous chapter we have discussed the problem of an ideal shock wave for which a jump in the field variables occurs across the shock surface. In reality if we look the experiments, we recognize that we have a very rapid but continuous passage from one to another equilibrium state. This is the well known problem of shock structure solution in which the shock have a *thickness*. From the theoretical point of view the problem of giving a satisfactory explanation about the behaviour of the thickness (of order of mean free path) of shock wave structure for increasing Mach numbers, using a continuum model, is an old and open question. Infact the classical parabolic models like Navier-Fourier or Burnett predict not satisfactory results [29] compared with the experiments [30], while the hyperbolic models as the 13-moments Grad equations or the equivalent continuum models of *Extended Thermodynamics* [24] do not admit continuous solution after a very small critical Mach number (*sub-shocks* formation) [31, 32]. In this last case, recently Jou and Pavon [33], taking into account that the previous hyperbolic models are linear in the non equilibrium variables as heat flux and shear stress, have conjectured, resuming a previous opinion of Anile and Majorana [32], that for increasing the critical Mach number it is necessary to add non linear terms in these variables. Another natural conjecture is that the critical Mach number increases and the thickness behaviour becomes more satisfactory if one considers in the Boltzmann equation more and more moments instead of the usual 13 moments. This idea is sustained by similar problems as the phase velocity in the limit of high frequencies or in the light scattering in which it was necessary to consider more moments than 13 to obtain a good agreement with the experimental data [35, 24] .

Now we present a recent result [34] proving that both the conjectures are not true. For this aim it is necessary to understand the mathematical general reasons of the breakdown of shock wave structure C^1 solutions for a generic hyperbolic dissipative system of balance laws to which the Extended Thermodynamics models belong. Then is very simple to deduce an upper bound such that for shock velocity greater than this limit no continuous shock wave structure solutions may exist. The first surprising result lies in the fact that this quantity is a characteristic of the associated linearized (in the neighbourhood of the equilibrium unperturbed state) equations. Therefore, in the case of Extended Thermodynamics this upper bound does not change adding non linear non-equilibrium terms. Moreover it coincides in the context of moments theory with the smallest characteristic eigenvalue of the linearized system that is greater than the sound velocity. Taking into account that Weiss [35] has evaluated explicitly all the characteristic eigenvalues of the linear system up to the remarkable number of 5456 moments, we are able to establish that our upper bound critical Mach number is not a monotonous function of the moments number but oscillates in the neighbourhood of 1 when the moments number increases. Therefore it is impossible to get good results adding non linear terms, or considering more and more moments. We conclude that the behaviour of the thickness as function of Mach number is not in the range of any continuum non-equilibrium thermodynamics compatible, in the sense of moments, with the Boltzmann equation. On the other hand, as we have seen in the previous chapter the system of Extended Thermodynamics is in the form of Balance Laws, and then it is always possible to study shocks without thickness like in the non dissipative case. The shock thickness remains only a microscopic phenomenon.

7 The Balance Laws System

Let rewrite a generic system of $N-$balance laws (1.3) in one space-dimension:

$$\frac{\partial \mathbf{F}(\mathbf{u})}{\partial t} + \frac{\partial \mathbf{G}(\mathbf{u})}{\partial x} = \mathbf{f}(\mathbf{u}) \tag{156}$$

where the *densities* $\mathbf{F}$, the *fluxes* $\mathbf{G}$ and the *productions* $\mathbf{f}$ are $\Re^N$-column vectors depending on the space variable x and time t through the field $\mathbf{u} \equiv \mathbf{u}(x,t) \in \Re^N$. In the physical examples, typically in the Extended Thermodynamics (6.1)-(6.5), $M < N$ equations of (156) represent *conservation laws.* Therefore, we suppose that the system (156) is equivalent to

$$\begin{cases} \dfrac{\partial \mathbf{V}(\mathbf{u})}{\partial t} + \dfrac{\partial \mathbf{P}(\mathbf{u})}{\partial x} & = 0 \\[2ex] \dfrac{\partial \mathbf{W}(\mathbf{u})}{\partial t} + \dfrac{\partial \mathbf{R}(\mathbf{u})}{\partial x} & = \mathbf{g}(\mathbf{u}) \end{cases} \tag{157}$$

where $\mathbf{V}, \mathbf{P} \in \Re^M$, while $\mathbf{W}, \mathbf{R}$ and $\mathbf{g}$ are vectors of $\Re^{N-M}$:

$$\mathbf{F} \equiv \begin{bmatrix} \mathbf{V} \\ \mathbf{W} \end{bmatrix}; \qquad \mathbf{G} \equiv \begin{bmatrix} \mathbf{P} \\ \mathbf{R} \end{bmatrix}; \qquad \mathbf{f} \equiv \begin{bmatrix} 0 \\ \mathbf{g} \end{bmatrix}.$$

Moreover, we assume that it is possible to consider the field $\mathbf{u}$ constituted by a pair $\mathbf{u} \equiv (\mathbf{v}, \mathbf{w})^T$ (the apex T denotes transpose) with $\mathbf{v} \in \Re^M$ and $\mathbf{w} \in \Re^{N-M}$, such that

$$\mathbf{g}(\mathbf{v}, 0) \equiv 0 \ \forall \mathbf{v}; \qquad \mathbf{g}(\mathbf{v}, \mathbf{w}) \neq 0 \ \forall \mathbf{v}, \ \forall \mathbf{w} \neq 0. \tag{158}$$

We call the generic state for which $\mathbf{w} = 0$ an *equilibrium state* and the $N - M$ components of $\mathbf{w}$ characterise the *non-equilibrium state variables.*
We require also that the system (156) is hyperbolic in the time direction and we indicate with $\lambda^{(k)}(\mathbf{u}) \quad (k = 1, 2, \ldots, N)$ the eigenvalues of $\partial \mathbf{G}(\mathbf{u})/\partial \mathbf{u}$ with respect to the matrix $\partial \mathbf{F}(\mathbf{u})/\partial \mathbf{u}$, i.e. solutions of

$$\det \left(\frac{\partial \mathbf{G}(\mathbf{u})}{\partial \mathbf{u}} - \lambda \frac{\partial \mathbf{F}(\mathbf{u})}{\partial \mathbf{u}} \right) \doteq 0.$$

The λ's are the characteristic velocities of the system (156) that, for the hyperbolicity assumption, are all real and finite. Moreover we call the system

$$\frac{\partial \mathbf{V}(\mathbf{v}, 0)}{\partial t} + \frac{\partial \mathbf{P}(\mathbf{v}, 0)}{\partial x} = 0, \tag{159}$$

obtained by the system $(157)_1$ putting identically equal to zero the non equilibrium variables $\mathbf{w}$, the *equilibrium sub-system* associated to the system (157) and we represent with $\mu^{(J)}(\mathbf{v}) \quad (J = 1, 2, \ldots, M)$ the corresponding characteristic velocities:

$$\det \left(\frac{\partial \mathbf{P}(\mathbf{v}, 0)}{\partial \mathbf{v}} - \mu \frac{\partial \mathbf{V}(\mathbf{v}, 0)}{\partial \mathbf{v}} \right) = 0. \tag{160}$$

Of course, no relation exists, in general, between the characteristic velocities $\mu^{(J)}(\mathbf{v})$ of the equilibrium sub-system and the equilibrium values of the characteristic velocities $\lambda^{(k)}(\mathbf{v},0)$ of the full system.

8 Shock Wave Structure

As it is well known, a shock wave structure is a regular solution of (156) depending on one variable

$$\mathbf{u} \equiv \mathbf{u}(\varphi); \qquad \varphi = x - st, \qquad s = \text{const.} \qquad \text{(shock velocity)} \tag{161}$$

and such that

$$\lim_{\varphi \to \pm\infty} \mathbf{u}(\varphi) = \begin{cases} \mathbf{u}_o \\ \mathbf{u}_1 \end{cases}; \qquad \lim_{\varphi \to \pm\infty} \frac{d\mathbf{u}}{d\varphi} = 0, \tag{162}$$

i.e. a wave solution connecting two constant states[6]. Substituting (161) into (156), we have an ordinary differential system

$$\left(-s\frac{\partial \mathbf{F}(\mathbf{u})}{\partial \mathbf{u}} + \frac{\partial \mathbf{G}(\mathbf{u})}{\partial \mathbf{u}}\right)\frac{d\mathbf{u}}{d\varphi} = \mathbf{f}(\mathbf{u}), \tag{163}$$

or, equivalently, from (157):

$$\begin{cases} \frac{d}{d\varphi}(-s\mathbf{V}(\mathbf{v},\mathbf{w}) + \mathbf{P}(\mathbf{v},\mathbf{w})) & = & 0, \\ -s\frac{d}{d\varphi}\mathbf{W}(\mathbf{v},\mathbf{w}) + \frac{d}{d\varphi}\mathbf{R}(\mathbf{v},\mathbf{w}) & = & \mathbf{g}(\mathbf{v},\mathbf{w}). \end{cases} \tag{164}$$

Taking into account $(162)_2$, we obtain from $(164)_2$ evaluated at $\varphi \to \pm\infty$

$$\mathbf{g}(\mathbf{v}_o, \mathbf{w}_o) = \mathbf{g}(\mathbf{v}_1, \mathbf{w}_1) = 0 \tag{165}$$

implying, from (158), that the solutions at infinity are equilibrium solutions:

$$\mathbf{w}_o = \mathbf{w}_1 = 0. \tag{166}$$

From $(164)_1$ the conservation along the process of

$$-s\mathbf{V}(\mathbf{v},\mathbf{w}) + \mathbf{P}(\mathbf{v},\mathbf{w}) = \mathbf{c} = \text{const.} \tag{167}$$

---

(6) In the usual physical cases the balance laws system satisfies the Galilean invariance. Therefore it is possible to consider the frame moving with the shock velocity and for this observer the wave appears stationary: $\varphi = x$. The interested reader may read the paper [36] in which the constraints for the system (156) due to the Galilean invariance are discussed in detail.

follows and so, in particular, for $\varphi \to \pm\infty$ (see (166))

$$\mathbf{c} = -s\mathbf{V}(\mathbf{v}_o, 0) + \mathbf{P}(\mathbf{v}_o, 0) = -s\mathbf{V}(\mathbf{v}_1, 0) + \mathbf{P}(\mathbf{v}_1, 0). \tag{168}$$

Equation (168) coincides with the Rankine-Hugoniot compatibility conditions (3.30) for shocks of the *equilibrium sub-system* (159).
From (168), except the trivial solution $\mathbf{v}_1 = \mathbf{v}_0$ (null shock) that there exists always for all s, it is possible for a fixed value of $\mathbf{v}_o$ (*unperturbed state*) to determine the *perturbed equilibrium state* $\mathbf{v}_1$ as a function of the shock parameter[(7)] s :

$$\mathbf{v}_1 \equiv \mathbf{v}_1(\mathbf{v}_o, s). \tag{169}$$

An *admissible* shock (169), solution of (168), must satisfy the Lax conditions; i.e. in correspondence with a fixed eigenvalue μ of the equilibrium sub-system (159) one has

$$\mu_o < s < \mu_1; \qquad \lim_{s \to \mu_o} \mathbf{v}_1(\mathbf{v}_o, s) = \mathbf{v}_o; \qquad (\mu_o = \mu(\mathbf{v}_o);\ \mu_1 = \mu(\mathbf{v}_1)). \tag{170}$$

For the following it is important to point out that, for the previous considerations, the shock velocity s satisfies the inequality:

$$s \geq \mu_o. \tag{171}$$

For a fixed value of $\mathbf{v}_o$ and s the constant vector $\mathbf{c}$ in (167) is determined by the first equality of (168) and therefore, at least in principle, it is possible to solve (167) in $\mathbf{v}$:

$$\mathbf{v} \equiv \mathbf{v}(\mathbf{w}, \mathbf{v}_o, s). \tag{172}$$

Inserting (172) in $(164)_2$ we obtain a non-linear ordinary differential system of $N-M$ equations (depending of $M+1$ parameters: s and $\mathbf{v}_o$) for the *non-equilibrium* vector $\mathbf{w} \in \Re^{N-M}$ as function of φ vanishing at $\pm\infty$.

9 Breakdown of the Solution

In this section, a simple theorem is proved permitting to determine an upper bound, easy to evaluate, such that for shock velocity greater than this limit C^1 shock wave structure solutions cannot exist.

Theorem - *We consider a system of balance laws* (157) *and suppose*

$$\max_{k=1,2,\dots,N} \lambda_o^{(k)} > \max_{J=1,2,\dots,M} \mu_o^{(J)}; \qquad (\lambda_o = \lambda(\mathbf{v}_o, 0);\ \mu_o = \mu(\mathbf{v}_o)), \tag{173}$$

(7) We exclude the exceptional case of characteristic shocks in which the shock velocity coincides identically with a characteristic eigenvalue as in the linear case (see for more details [37]).

where $\mathbf{u}_o \equiv (\mathbf{v}_o, 0)^T$ *is the unperturbed equilibrium state of a shock wave structure. For a prefixed eigenvalue* $\mu_o \in \mu_o^{(J)}$, $J = 1, 2, ..., M$ *of the equilibrium system, we start from the trivial shock for which* $s = \mu_o$ *and we increase* s *satisfying the Lax condition* $s > \mu_o$.
Let $\widetilde{\lambda}(\mathbf{u})$ *the smallest* $\lambda^{(k)}(\mathbf{u})$ *greater than* μ_o *and* $\widetilde{\mathbf{l}}(\mathbf{u})$, $\widetilde{\mathbf{d}}(\mathbf{u})$ *the corresponding left and right caracteristic eigenvectors:*

$$\widetilde{\mathbf{l}} \cdot \left(\frac{\partial \mathbf{G}(\mathbf{u})}{\partial \mathbf{u}} - \widetilde{\lambda} \frac{\partial \mathbf{F}(\mathbf{u})}{\partial \mathbf{u}} \right) = 0, \quad \left(\frac{\partial \mathbf{G}(\mathbf{u})}{\partial \mathbf{u}} - \widetilde{\lambda} \frac{\partial \mathbf{F}(\mathbf{u})}{\partial \mathbf{u}} \right) \cdot \widetilde{\mathbf{d}} = 0. \tag{174}$$

Let us assume that the characteristic velocity $\widetilde{\lambda}$ *satisfies the genuine non linearity condition:*

$$\nabla \widetilde{\lambda} \cdot \widetilde{\mathbf{d}} \neq 0, \tag{175}$$

and the production term $\mathbf{f}$ *verifies the following dissipative conditions* [8]:

$$\widetilde{\mathbf{l}} \cdot \mathbf{f} \neq 0 \qquad \forall \mathbf{w} \neq 0; \tag{176}$$

$$\left(\widetilde{\mathbf{l}} \cdot \nabla \mathbf{f} \cdot \widetilde{\mathbf{d}} \right)_o \neq 0. \tag{177}$$

Then under the conditions (173), (175), (176) *and* (177), *there exists always a finite critical value* s_c

$$\mu_o < s_c \leq \widetilde{\lambda}_o \tag{178}$$

of the shock velocity such that a breakdown of the C^1 *shock structure solution happens. In particular, it is impossible to have a* C^1 *solution for* $s \geq \widetilde{\lambda}_o$.

Proof: Starting from the null shock $s = \mu_o$, let's now increase s and suppose that, for a given $s_* > \mu_o$ there exists still a C^1 solution $\mathbf{u}_*(\varphi)$ of (163) connecting the two equilibrium states $\mathbf{u}_o \equiv (\mathbf{v}_o, 0)^T$ and $\mathbf{u}_1 \equiv (\mathbf{v}_1(\mathbf{v}_o, s_*), 0)^T$. We multiply (163) by $\widetilde{\mathbf{l}}(\mathbf{u}_*)$ obtaining from (174):

$$\widetilde{\mathbf{l}}(\mathbf{u}_*) \cdot \frac{\partial \mathbf{F}}{\partial \mathbf{u}}(\mathbf{u}_*) \frac{d\mathbf{u}_*}{d\varphi} = \frac{\widetilde{\mathbf{l}}(\mathbf{u}_*) \cdot \mathbf{f}(\mathbf{u}_*)}{\widetilde{\lambda}(\mathbf{u}_*) - s_*}, \tag{179}$$

and taking into account the condition (176) we must have:

$$\mu_o \leq s_* < \widetilde{\lambda}(\mathbf{u}_*) \tag{180}$$

at least for all $\varphi \in]-\infty, +\infty[$ for which the production $\mathbf{f}$ does not vanish. The condition (180) is surely verified also in the limit $\varphi \to +\infty$. In fact, if ad absurdum $s_* = \widetilde{\lambda}_o$ since at $+\infty$ the production $\mathbf{f}$ vanishes, the second member of (179) is undetermined. The application of De l'Hospital theorem permits to evaluate in this circumstance the second member of (179):

$$\left(\frac{\widetilde{\mathbf{l}}\cdot\nabla\mathbf{f}\cdot\frac{d\mathbf{u}_*}{d\varphi}}{\nabla\widetilde{\lambda}\cdot\frac{d\mathbf{u}_*}{d\varphi}}\right)_o .$$

Taking into account from (163) that in this case $(d\mathbf{u}_*/d\varphi)_o$ is proportional to the right eigenvector $\widetilde{\mathbf{d}}_o$ and on the assumptions (177) and (175) we obtain that the second member is not zero in contrast with the boundary conditons $(162)_2$. Therefore we conclude that if there exists a C^1 solution then the condition (180) results to be verified for all φ and therefore in particular ome has:

$$\mu_o \leq s < \widetilde{\lambda}_o, \tag{181}$$

and it is impossible to have a C^1 solution when $s \geq \widetilde{\lambda}_o$.

- Remark 1: As it is well known the characteristic eigenvalues evaluated at equilibrium λ_o, and in particular $\widetilde{\lambda}_o$, coincide with the characteristic velocities of the linear system obtained linearizing the system (1) in the neighbourhood of the equilibrium state $\mathbf{u}_o$. They also are coincident with the phase velocities of the linearized problem in the limit of high frequency (see e.g. [8]). Therefore since our problem is fully non linear, the critical shock velocity $s_c\,(\mu_o < s_c \leq \widetilde{\lambda}_o)$ may depend on the non equilibrium fields but, according to our result, its maximum value $\widetilde{\lambda}_o$ *results on the contrary independent i.e.* $\widetilde{\lambda}_o$ *does not change modifying or adding non linear non equilibrium terms in* $\mathbf{w}$.

- Remark 2: We observe that the only exceptional case in which an upper critical limit is unknown occurs when the condition (173) is violated (this circumstance is never verified until now in the physical applications) and we choose as μ_o the greatest of the μ's.

10 Breakdown in the Extended Thermodynamics

This example (6.1)-(6.5) belongs to the previous general framework having now

$$N = 5,\ M = 3,\ \mathbf{v} \equiv (\rho, v, \theta)^T,\ \mathbf{w} \equiv (q, \sigma)^T. \tag{182}$$

The two equilibrium states $\mathbf{u}_o$ (unperturbed) and $\mathbf{u}_1$ (perturbed) are in according to the condition (166):

$$\mathbf{u}_o \equiv (\rho_o, v_o = 0, \theta_o, q_o = 0, \sigma_o = 0)^T; \quad \mathbf{u}_1 \equiv (\rho_1, v_1, \theta_1, q_1 = 0, \sigma_1 = 0)^T. \tag{183}$$

The values ρ_1, θ_1 and v_1 are related through the well known (see e.g. [28]) Rankine-Hugoniot equations (168) to ρ_o, θ_o and depend on the shock parameter $M_o = s/c_o$ (with $c_o = \sqrt{(5/3)k\theta_o}$ denoting the sound velocity).
In this case the equilibrium sub-system (159) coincides with the equations (6.1)-(6.3) when $q \equiv 0$ and $\sigma \equiv 0$, i.e. with the Euler equations and so:

$$\mu_o^{(1)} = -c_o; \quad \mu_o^{(2)} = 0; \quad \mu_o^{(3)} = c_o. \tag{184}$$

Instead, the equilibrium characteristic eigenvalues of the full system are [24]:

$$\lambda_o^{(1)} = -1.65c_o;\ \lambda_o^{(2)} = -0.62c_o;\ \lambda_o^{(3)} = 0;\ \lambda_o^{(4)} = 0.62c_o;\ \lambda_o^{(5)} = 1.65c_o. \tag{185}$$

Choosing the shock travelling in the $x > 0$ direction, we have

$$\mu_o = c_o \quad \text{and} \quad \tilde{\lambda}_o = \lambda_o^{(5)} = 1.65c_o, \tag{186}$$

and therefore C^1 solutions can exist only for Mach numbers such that:

$$1 \leq M_o < 1.65 \tag{187}$$

while for

$$M_o \geq 1.65 \tag{188}$$

C^1 solutions are forbidden. So we find again, in an immediate way, the well known Grad's result [31, 32]. It is interesting to note that the critical Mach number determined by Grad [31] and Anile & Majorana [32] through an heavy numerical integration of the system (163) coincides in this case with our upper limit (188).
Besides, as consequence of the remark n.1 of the previous section, also if the critical Mach number may depend on the non equilibrium variables, it is impossible to change its upper limit (188) adding non linear terms in q or σ, in contrast therefore with one of the conjectures of [33].

11 More and More Moments

A natural question is to ask what happens by considering more moments than 13. In this case the conservation laws remain always the usual ones and therefore μ_o is still equal to c_o, the λ_o's changing every time we increase the moments number. Weiss [35], studying the problem of the behaviour of phase velocity in the limit of high frequency has evaluated all the λ_o's until 5456 moments: therefore it is a simple matter to establish, for a fixed number of moments, the minimum of eigenvalues greater than $\mu_o = c_o$. In figure 9 we plot our upper bound $\tilde{\lambda}_o/c_o$ for increasing moments (take into account that 5456 moments correspond, in the one dimensional case, to 256 equations). From figure 9 it is evident that $\tilde{\lambda}_o/c_o$ oscillates with a small damping over 1. Therefore, if the moments number increases, we have always a breakdown of the solution for critical Mach numbers a little more greater than one. Then, we conclude that in a continuum approach, independently by the variables number, it is impossible to find C^1 solutions and only shocks without thickness discussed in the previous chapter have validity according with the assumption that in the continuum limit the mean free path tends to zero.

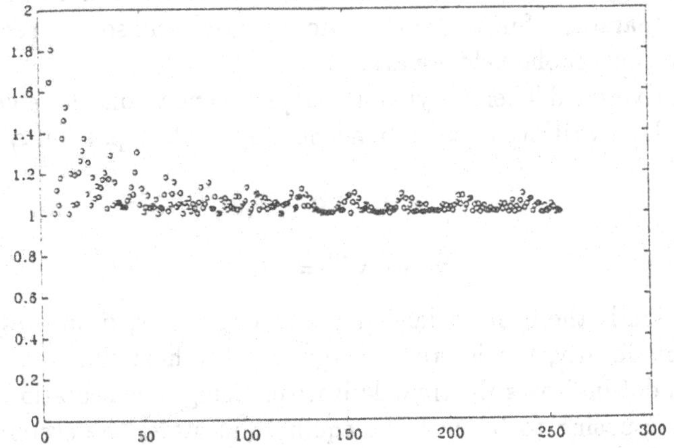

fig.9

IV. Phonon Gas and Changes of Shape of Second Sound Wave

In 1947 Peshkow [39] suggested that heat could propagate in pure crystals as a true temperature wave, called second sound. In the following years a great work has been developed to understand the theoretical bases of this idea (see, in particular, the papers of Guyer and Krumhansl [40], [41]) and for finding experimentally the new wave. At the first time, second sound was observed in pure crystals of ^4He (1966) and then in high-purity crystals of ^3He (1969), NaF (1970) and Bi (1972). To study the heat pulses in very pure crystals at low temperatures the starting point lies in considering the crystal as a phonon system. Here the normal processes (N-processes) in which phonon momentum is conserved are stronger, in certain temperature ranges, than the R-processes (dissipative processes not conserving momentum) and so the second sound can be identified. Two interesting features must be underlined: the first is the existence of a *critical temperature* such that the second sound is most clearly seen (for example, about 15^oK in NaF and 3.5^oK in Bi) and the second one concerns the modifications of the initial square wave form during its propagation according to the different temperatures of the crystal.

12 Generalized Maxwell-Cattaneo equation and second sound propagation

The phenomenology previously illustrated cannot be interpreted by Fourier's theory because of the "paradox of instantaneous propagation" and so it is necessary to find a suitable set of hyperbolic field equations.
In the spirit of Extended Thermodynamics [24] let us now consider a general system of two balance laws writing, in correspondence to the state pair $(\theta, \mathbf{q})$

$$\rho\dot{e} + \operatorname{div}\mathbf{q} = 0 \tag{189}$$

$$\dot{\mathbf{w}} + \operatorname{div}\mathbf{T} = -\mathbf{b}. \tag{190}$$

The first equation is the usual balance law of energy; ρ, e, $\mathbf{q}$, are respectively the (constant) mass density, the internal energy and the heat flux vector. Moreover the superposed dot indicates the time derivative. Using representation theorems for $\mathbf{w}$, $\mathbf{T}$, $\mathbf{b}$ and supposing to be near the equilibrium state, the system (189)-(190) becomes [7], [42]-[44]

$$\rho\dot{e} + \operatorname{div}\mathbf{q} = 0 \tag{191}$$

$$(\alpha\mathbf{q})^{\cdot} + \nabla\nu = -\frac{\nu'}{\kappa}\mathbf{q} \tag{192}$$

Here $\kappa \equiv \kappa(\theta)$ represents the heat conductivity and the remaining functions α, ν together with e are constitutive quantities depending on the absolute temperature θ (the apex denotes the derivative with respect to θ and ∇ is the gradient operator). When α is equal to a constant, the Maxwell-Cattaneo equation

$$\tau\dot{\mathbf{q}} + \mathbf{q} = -\kappa\nabla\theta \tag{193}$$

is obtained ($\tau = \alpha\kappa/\nu'$) while, if $\alpha = 0$, we have the Fourier law.
The most important feature of (192) lies in the presence of the not constant factor α playing the role of thermal inertia. In fact, if $\alpha \equiv \alpha(\theta)$, the entropy principle as well as the hyperbolicity of the differential system (191)-(192) are satisfied without requiring the dependence of e on $\mathbf{q}$ in addition to temperature [45]. Besides the great generality due to the function $\alpha(\theta)$ allows us to recover the stability criterion of the maximum of entropy at equilibrium.
Let us impose now the compatibility of eqs. (191)-(192) with the entropy principle taken in the form

$$\dot{h}^{o} + \operatorname{div}\mathbf{h} \le 0 \tag{194}$$

with

$$h^o = -\rho S, \qquad \mathbf{h} = -\frac{\mathbf{q}}{\theta} \tag{195}$$

(S is the specific entropy). Then we obtain [44]

$$\alpha = \gamma/(\nu'\theta^2), \quad \gamma = \text{const.}, \qquad \kappa > 0, \tag{196}$$

$$h^o = -\rho S = -\rho S_E(\theta) + \frac{\gamma q^2}{2(\nu'\theta^2)^2} \tag{197}$$

where S_E is the equilibrium entropy density.
Also the convexity condition for h^o, with respect to the field $\mathbf{u} \equiv (\rho e, \alpha \mathbf{q})^T$ is imposed and this implies our system is symmetric-hyperbolic (in the sense of Friedrichs) if

$$\gamma > 0, \qquad c(\theta) = e'(\theta) > 0 \tag{198}$$

where c is the equilibrium specific heat.
Taking into account that $e = e(\theta)$ is known (for example, in the case of crystals at low temperature $e = \epsilon\theta^4/4$) and also $\kappa(\theta)$ is found through experimental data, we have at this step that the only arbitrary quantities are $\nu(\theta)$ and the constant γ. Besides the second sound velocity at equilibrium, $U_E \equiv U_E(\theta)$, can be identified with the characteristic velocities of the system (191)-(192) evaluated in an equilibrium state ($\mathbf{q} = 0$). The characteristic velocities in a generic state are given by the roots of the characteristic polynomial

$$\rho c \alpha \lambda^2 + \lambda \alpha' q_n - \nu' = 0, \tag{199}$$

where $q_n \equiv \mathbf{q} \cdot \mathbf{n}$ and $\mathbf{n}$ is the unit normal to the characteristic wave front.
Therefore from $(196)_1$ and (199), when $\mathbf{q} = 0$, the constitutive function ν in terms of U_E is obtained

$$\frac{\nu}{\sqrt{\rho\gamma}} = \int \frac{U_E(\theta)}{\theta}\sqrt{c(\theta)}d\theta. \tag{200}$$

Since it is possible to verify that γ is an inessential common factor we have no more free parameters: in other words all the constitutive functions are univocally determined knowing the equilibrium quantities $e \equiv e(\theta)$, $\kappa \equiv \kappa(\theta)$, $U_E \equiv U_E(\theta)$ [44].

13 Shock waves in high purity crystals

As (191), (192) represent a system of balance laws (i.e. the first member is in the form of space-time divergence), it is possible to write it in an integral form and to study weak solutions and, in particular, shock waves [17]; then the Rankine-Hugoniot compatibility conditions across the shock front allow us to evaluate the shock velocity s in terms of the temperature θ_o, θ_1 respectively ahead and behind the shock surface.

Let us apply now the selection rules of the entropy growth and the Lax conditions to the present approach in the case of NaF and Bi crystals specifying only the functions $U_E(\theta)$ and $e(\theta)$. The values of U_E obtained from experiments by Jackson et al. (for NaF) [47] and by Narayanamurti-Dynes (for Bi) [48] are well described by the empirical equation [49]

$$U_E^{-2} = A + B\theta^n \tag{201}$$

in the temperature range $10^oK \leq \theta \leq 18.5^oK$ (for NaF) and $1.4^oK \leq \theta \leq 4^oK$ (for Bi), where heat pulses were observed with properties expected of second sound. Values of the parameters A, B, n giving an excellent fit are [49]

$$n = 3.10, \quad A = 9.09 \cdot 10^{-12}, \quad B = 2.22 \cdot 10^{-15} \quad \text{(NaF)}$$

$$n = 3.75, \quad A = 9.07 \cdot 10^{-11}, \quad B = 7.58 \cdot 10^{-13} \quad \text{(Bi)}$$

for U_E in centimeters per second and θ in Kelvin degrees. Furthermore we take the equilibrium specific heat $c = \epsilon\theta^3$, with $\epsilon = 23$ erg cm^{-3} $^oK^{-4}$ for NaF and $\epsilon = 550$ erg cm^{-3} $^oK^{-4}$ for Bi.

Considering a plane shock wave propagating in the x-direction ($\mathbf{n} \equiv (1,0,0)$), from the Rankine-Hugoniot equations it is possible to obtain $s = s(\theta_o, \theta_1)$ and $\lambda_1 = \lambda_1(\theta_o, \theta_1)$ where θ_o is the unperturbed temperature while θ_1 is the perturbed one (shock parameter).

Then, a numerical evaluation [17] allows us to plot s and λ_1 vs. temperature θ_1, for a fixed value of θ_o, in both cases of NaF and Bi. Figures $10 \div 12$ refer to NaF case. We observe that when the temperature θ_o increases in the range $10^oK \sim 18.5^oK$ the plots are, at first, of type displayed in fig. 10 and then as in fig. 11.

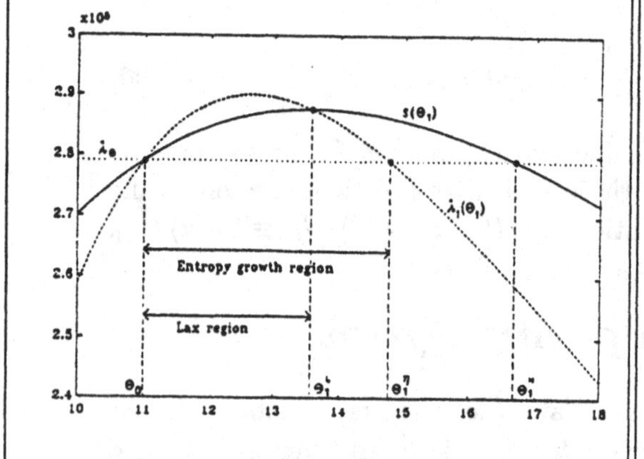

fig.10

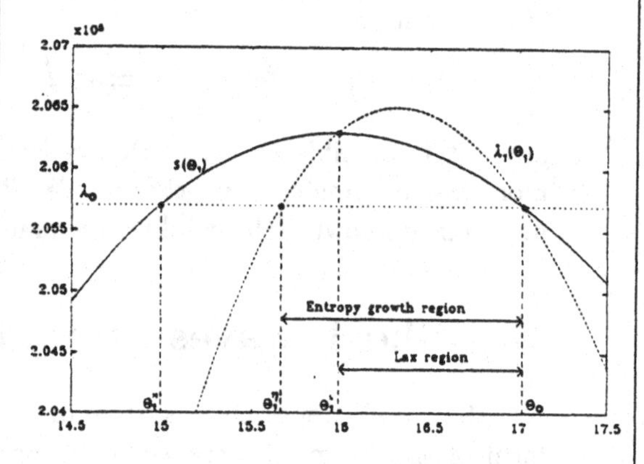

fig.11

Note that, in both figures, the Lax conditions impose that the possible shocks there exist only if $|\theta_1 - \theta_o|$ is bounded (unlike the usual shocks which occur, for example, in fluid dynamics). In particular in fig. 10 it is clearly seen that the Lax conditions are verified in the range $\theta_o < \theta_1 < \theta_1{}^L$ ($\theta_1{}^L$ depending on θ_o). The shock wave can then propagate through the material only if we generate a heat pulse with a positive jump of temperature not exceeding the maximum value of $\theta_1{}^L - \theta_o$. Let us call this shock a *hot shock*. In fig. 11 we note that there is a very different physical situation since the Lax conditions are verified in the range $\theta_1{}^L < \theta_1 < \theta_o$. The shock propagation takes place now if the initial temperature jump is negative and does not exceed in absolute value $|\theta_1{}^L - \theta_o|$ (*cold shock*).

The transition from a situation to the other one is shown in fig. 12 where it is pointed out the existence of a *critical temperature* $\tilde{\theta} = 15.36°K$ such that $\theta_o = \tilde{\theta} = \theta_1{}^L$. In this particular case, the Lax conditions are not satisfied and no shock is possible.

It turns out that $\tilde{\theta}$ is a structural temperature, i.e. characteristic of NaF, defining the boundary between two very different phenomena: for $\theta_o < \tilde{\theta}$ a hot shock is generated while the cold shock appears for $\theta_o > \tilde{\theta}$ and we point out that $\tilde{\theta}$ is the temperature for which the heat flux behind the front changes sign [17].

The same qualitative behaviour is observed in Bi in the range $1.4°K \sim 4°K$. In this case the critical temperature $\tilde{\theta} = 3.38°K$ is found.

The non usual *cold shock* might appear inconsistent with thermodynamics but the study of the function η characterizing the entropy growth across the shock surface shows that $\eta > 0$ in the Lax region. Furthermore note that the temperature range for which $\eta > 0$ is larger than the previous one: $\theta_1{}^\eta < \theta_1{}^L < \theta_1 < \theta_o$ (in fig. 13 η/ρ vs. θ_1 with a fixed $\theta_o > \tilde{\theta}$, i.e. in the case of cold shock, is plotted for NaF).

The condition $\eta > 0$ provides the same qualitative results of the Lax conditions also for hot shocks with a $\theta_1{}^\eta > \theta_1{}^L$. Observing as the plot of the function η/ρ is modified changing the unperturbed temperature θ_o, it results that the value of the critical temperature $\tilde{\theta}$ remains unchanged.

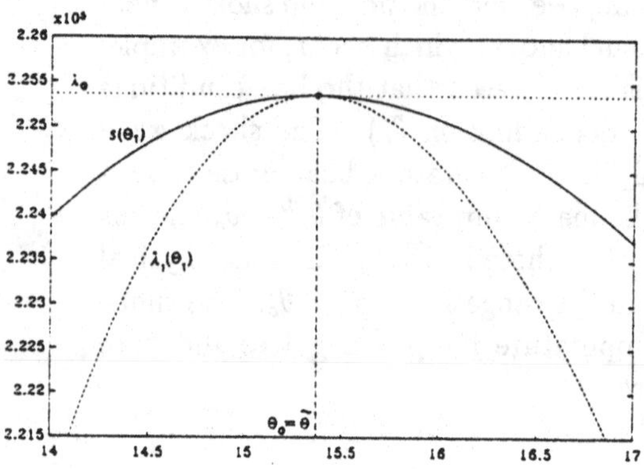

fig.12

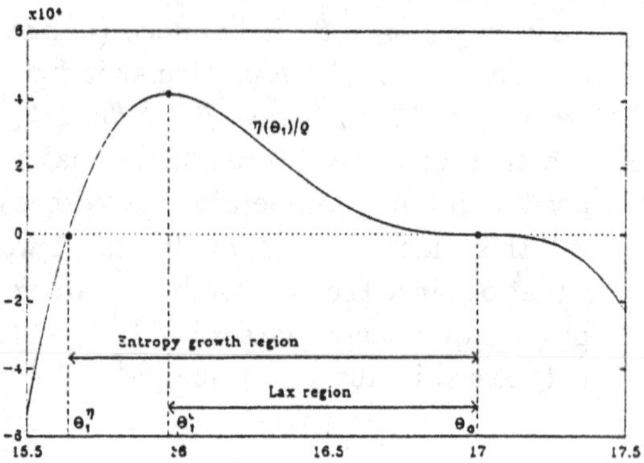

fig.13

So we want to remark that the present model shows unusual shocks characterized, from a macroscopic point of view, by the existence of a critical structural temperature $\tilde{\theta}$ for which the "state" of the material changes in a very unexpected way. In particular the value $\tilde{\theta} = 15.36^oK$ is found very close to the value ($\sim 15^oK$) at which a new pulse, identified as second sound, is clearly seen in a highly pure dielectric crystal of NaF. Also the value $\tilde{\theta} = 3.38^oK$ is practically coincident with the value ($\sim 3.5^oK$) at which the saturation of the velocity in the second sound regime has been observed in a pure crystal of the semimetal bismuth.
The value $\tilde{\theta}$ of the critical temperature was found by the plots of s and λ_1 changing the value of θ_o. In fact, the presence of two very different shocks (hot and cold shocks) enables us to find numerically the transition temperature for which any shock at all is forbidden by the Lax conditions.
However using the bifurcation analysis of the Rankine Hugoniot equations described in Part I, it is possible to find the particular unperturbed state $\tilde{\mathbf{u}}_o$, such that the Lax conditions are violated also for weak shocks, i.e.

$$(\nabla\lambda \cdot \mathbf{d})_{\tilde{\mathbf{u}}_o} = 0 \tag{202}$$

In the present case (202) implies that $\tilde{\theta}$ is the value for which the function

$$\Phi(\theta) = U_E(\theta)\theta^{5/6} \tag{203}$$

has a maximum. Using for U_E the empirical relationship (201) it follows

$$\tilde{\theta} = \left\{\frac{5A}{B(3n-5)}\right\}^{1/n} \tag{204}$$

and then also $\tilde{U}_E$ can be found analytically i.e.

$$\tilde{U}_E = \sqrt{\frac{3n-5}{3nA}}. \tag{205}$$

The relationships (204), (205) give

$$\tilde{\theta} = 15.36^o K, \qquad \tilde{U}_E = 2.26 \cdot 10^5 \text{cm/sec} \qquad \text{(for NaF)} \tag{206}$$

$$\tilde{\theta} = 3.38^o K, \qquad \tilde{U}_E = 7.83 \cdot 10^4 \text{cm/sec} \qquad \text{(for Bi)} \tag{207}$$

It is interesting to underline that using the function (203) it is possible to find $\tilde{\theta}$ also for the cases of ^3He and ^4He; as it will be reported in [17], the values so obtained are again very close to the values for which the second sound is clearly picked out in these crystals.

14 Changes of shape on second sound wave

To conclude we present a possible explanation, based on the previous general results, of the distortion of the initial thermal pulse during its propagation in a rigid heat conductor. The results obtained could be a check for verifying experimentally the limits of validity of our model. Suppose we generate an heat pulse by some type of heater: usually, with a good approximation, the schematic shape of the initial pulse is rectangular. It is possible to imagine the rectangular profile of the initial wave as two successive shock fronts: the first one is generated when the heater is on and corresponds to the hot shock studied in the previous section; therefore this shock is stable and can be propagated if $\theta_o < \tilde{\theta}$ and $\theta_1 < \theta_1^L$. The second one (heater off), does not correspond to the cold shock, because at present the right side of the shock (unperturbed region) is the non equilibrium state $(\theta_1, q_1 \neq 0)$, while the left side (perturbed region) is the equilibrium state $(\theta_o, q_o = 0)$. Since in the previous shock analysis we have chosen always the right side coincident with the equilibrium state, then to study the second shock it is necessary to change the equilibrium state with the non equilibrium one and vice versa. However it is a simple matter, to show that the expressions of $s(\theta_o, \theta_1)$ and $\lambda_1(\theta_o, \theta_1)$ remain unchanged, while the entropy production across the second shock coincides, except for the sign, with the entropy production of the first shock. The Lax conditions become now, the complementary ones:

$$\lambda_1 < s < \lambda_o; \tag{208}$$

and therefore it is possible to use all the previous figures by considering as admissible region for the Lax conditions the complementary set in which (208) holds. Moreover, the entropy growth condition differs by the sign with respect the entropy growth condition of the first front.

Let's consider now the two cases: i) $\theta_o < \tilde{\theta}$; ii) $\theta_o \geq \tilde{\theta}$.

- Case i): $\theta_o < \tilde{\theta}$ (fig.10 for NaF)

  - i1) If $\theta_1 < \theta_1^L$, the Lax conditions are satisfied as regards the first front, but are violated for the second one and therefore the second front is unstable and the back part of the signal is regularised. This case corresponds, substantially, to *weak amplitude*.

  - i2) If $\theta_1 \geq \theta_1^L$ both the shocks violate the Lax conditions and the only possibility is a regularization of both sides of the wave (*strong amplitude*). Incidentally, we observe that when $\theta_1 > \theta_1^H$ the second front satisfies (208) but the proper Lax conditions are violated because the shock does not pass through the null shock.

- Case ii): $\theta_o \geq \tilde{\theta}$ (fig 11 for NaF)

  In this case, the first shock is always forbidden and the second one verifies, $\forall \theta_1 > \theta_o$, the Lax conditions (208): we have a shape change in which only the front part is regularized as independently of the amplitude.

The same results arise by using the entropy growth criteria.

This behaviour seems in a good agreement with all the available experiments. In particular in all the materials under consideration there exists a temperature range in a neighbourhood of $\tilde{\theta}$ for which the wave appears regularised; the second sound shape described in the case i1) it is well supported by the oscilloscope traces in the case of ^4He and, moreover, in the case of *NaF* it can be deduced indirectly observing that the difference between the arrival times for leading edges and peaks increases with the temperature. On the other hand (unlike the case of superfluid Helium) the shape as described in ii) does not appear so evident in the experiments. In our opinions the reason, probably, could be that when $\theta_o > \tilde{\theta}$ we have the transition from the second sound range to the diffusion one and so our hyperbolic model loses its validity and it is necessary to add parabolic terms. If this is true the critical temperature $\tilde{\theta}$ becomes the transition temperature separating the hyperbolicity region (second sound wave) from the parabolic region (diffusion).

References

[1] **G. Boillat**, *La Propagation des Ondes.* Gauthier-Villars, Parigi (1965).

[2] **G. Boillat**, *Chocs caractéristiques.* C.R. Acad. Sc. Paris **274A**, 1018 (1972).

[3] **S. K. Godunov**, *An interesting class of quasilinear systems.* Sov.Math. **2** (1961).

[4] **K.O. Friedrichs & P.D. Lax**, *Systems of conservation equations with a convex extension.* Proc. Nat. Acad. Sci. USA., **68** (1971).

[5] **G. Boillat**, *Sur l' éxistence et la recherche d'équations de conservation supplémentaires pour les systémes hyperboliques.* C.R.Acad.Sci., Paris, **278** A (1974).

[6] **T. Ruggeri & A. Strumia**, *Main field and convex covariant density for quasi-linear hyperbolic systems. Relativistic fluid dynamics.* Ann.Inst. H.Poincaré, **34** (1981).

[7] **T. Ruggeri**, *Struttura dei Sistemi alle derivate parziali compatibili con un Principio di Entropia e Termodinamica Estesa.* Suppl. BUMI - Fisica Matematica, **4**, 261 (1985).

[8] **A.Muracchini, T. Ruggeri** and **L. Seccia**, *Dispersion relation in High Frequency limit and non linear Wave Stability for Hyperbolic Dissipative System.* Wave Motion, **15**, (2) (1992).

[9] **G. Boillat & T. Ruggeri**, *On the evolution law of the weak discontinuities for hyperbolic quasi-linear systems.* Wave Motion **1**, 149 (1979).

[10] **T. Ruggeri**, *Stability and Discontinuity Waves for Symmetric Hyperbolic Systems*, in *Non linear Wave Motion.* Ed. by A. Jeffrey, Longman (1989).

[11] **P. Chen**, *Growth and decay of waves in solids* in Mechanics of Solids III, Handbuch der Physik, **6A/3**, 303 Springer-Verlag (1973).

[12] **Y. Choquet-Bruhat**, *Ondes Asymptotiques et Approchées pour des Systémes d'équation aux derivées partielles non linéaires*, J. Math. Pures et Appl. **48**, 117 (1969).

[13] **G. Boillat**, *Ondes asymptotiques non linéaires*, Ann. Mat. Pura Appl. **111**, 31 (1976).

[14] **K. S. Eckhoff**, *On Stability for Symmetric Hyperbolic Systems*, Journ. Diff. Eq. **40**, 94 (1981).

[15] **A. Donato & F. Oliveri**, *Instability Conditions for Symmetric Quasi Linear Hyperbolic Systems*, Atti Sem. Mat. Fis. Univ. Modena. **35**, 191 (1987).

[16] **P.D. Lax**, *Hyperbolic Systems of Conservation Laws.* Comm. Pure Appl. Math. **10**, 537 (1957);

[17] **T. Ruggeri**, **A.Muracchini** and **L. Seccia**, *Continuum Approach to Phonon Gas and Shape of Second Sound via Shock Waves Theory* Nuovo Cimento **16**, n.1, 15 (1994). See also *Shock Waves and Second Sound in a Rigid Heat Conductor: A Critical Temperature for NaF and Bi.* Phys. Rev. Lett. **64**, 2640, (1990).

[18] **G. Boillat & T. Ruggeri**, *Reflection and transmission of discontinuity waves through a shock wave. General theory including also the case of characteristic shocks.* Proc. of the Royal Soc. of Edinburgh., **83 A**, 17 (1979).

[19] **L. Brun**, *Ondes de choc finies dans les solides èlastiques*, in *Mechanical waves in solids*, J. Mandel and L. Brun Eds. Springer, Wien (1975).

[20] **A. Jeffrey**, *Quasilinear hyperbolic systems and waves* Pitman, London (1976).

[21] **P.D. Lax**, *Shock Waves and Entropy*, in *Contribution to non linear functional analysis*; Zarantonello ed. Academic Press, New York (1971).

[22] **G. Boillat**, *Sur une fonction croissant comme l'entropie et generatrice des chocs dans les systemes hyperboliques.* C.R. Acad. Sc. Paris **283**-A, 409 (1976).

[23] **C. Dafermos**, *Generalized characteristics in Hyperbolic Systems of conservation laws.* Arch. Rat. Mech. and Anal. **107**, 127 (1989).

[24] **I. Müller & T. Ruggeri**, *Extended Thermodynamics*, Springer Tracts on Natural Philosophy **37** - Springer Verlag - New York (1993).

[25] **T. Ruggeri & L. Seccia**, *Hyperbolicity and Wave propagation in Extended Thermodynamics.* Meccanica **24**, 127 (1989).

[26] **A. Muracchini & L. Seccia**, *Thermo-acceleration waves and shock formation in Extended Thermodynamics of gravitational gases.* Continuum Mech. Thermodyn. **1**, 227 (1989).

[27] **T. Ruggeri**, *Shock Waves in Hyperbolic Dissipative Systems: Non Equilibrium Gases.* Pitmàn Research Notes in Mathematics **227**. D. Fusco & A. Jeffrey Eds. Longman (1991).

[28] **L. Landau & E. Lifsits**, *Mècanique des Fluides*, pag. 418 Moscow: MIR (1971).

[29] **D. Gilbarg & D. Paolucci**, *The Structure of Shock Waves in the Continuum Theory of Fluids.* Journ. Rat. Mech. Anal. **2**, 617 (1953).

[30] **H. Alsemeyer**, *Density profiles in argon and nitrogen shock waves measured by the absorption of an electron beam.* J. Fluid Mech. **74**, 497 (1976).

[31] **H. Grad**, *The Profile of a Steady Plane Shock Wave.* Comm. Pure Appl. Math. **5**, 257 (1952).

[32] **A. M. Anile & A. Majorana**, *Shock structure for heat conducting and viscid fluids.* Meccanica **16** (3),149 (1981).

[33] **D. Jou & D. Pavon**, *Non local and Nonlinear effects in shock waves.* Phys. Rev. A **44** (10), 6496 (1991).

[34] **T. Ruggeri**, *Breakdown of Shock Wave Structure Solutions.* Phys. Rev. **47**-E, (6) (1993).

[35] **W. Weiss**, *Hierarchie der Erweiterten Thermodynamik.* Dissertation TU Berlin (1990).

[36] **T. Ruggeri**, *Galilean Invariance and Entropy Principle for Systems of Balance Laws.* Cont. Mech. Thermodyn. **1** (1989).

[37] **G. Boillat & T. Ruggeri**, *Characteristic shocks: completely and strictly exceptional systems.* Boll. Unione Mat. Ital. (U.M.I.) **15** A, 197 (1978).

[38] **N. Kopell & L.N. Howard**, *Bifurcations and Trajectories joining Critical Points.* Advances in Math. **18**, 306 (1975).

[39] **V. Peshkov**, in: *Report on an International Conference on Fundamental Particles and Low Temperature Physics* Vol. II, The Physical Society of London, (1947).

[40] **R. A. Guyer & J. A. Krumhansl**, *Solution of the Linearized Phonon Boltzmann Equation.* Phys. Rev. **148**, 766 (1966).

[41] **R. A. Guyer & J. A. Krumhansl,** *Thermal Conductivity, Second Sound, and Phonon Hydrodynamic Phenomena in Nonmetallic Crystals.* Phys. Rev. **148**, 778 (1966).

[42] **T. Ruggeri,** *Thermodynamics and Symmetric Hyperbolic Systems.* Rend. Sem. Mat. Univ. Torino. Fascicolo speciale *Hyperbolic Equations,* **167** (1987).

[43] **A. Morro & T. Ruggeri,** *Second Sound And Internal Energy In Solids.* Int. J. Non-Linear Mech., **22**, 27 (1987).

[44] **A. Morro & T. Ruggeri,** *Non-equilibrium properties of solids obtained from second-sound measurements.* J. Phys. C: Solid State Phys., **21**, 1743 (1988).

[45] **B. D. Coleman, M. Fabrizio & D. R. Owen,** *On the thermodynamics of second sound in dielectric crystals.* Arch. Rat. Mech. Anal. **80**, 135 (1982).

[46] **G. Boillat & T. Ruggeri,** *Limite de la vitesse des chocs dans les champs a densité d'energie convexe.* Compt. Rend. Acad. Sci. Paris, **289** A, 257 (1979).

[47] **H. E. Jackson, C. T. Walker & T. F. McNelly,** *Second Sound In NaF.* Phys. Rev. Lett. **25**, 26 (1970).

[48] **V. Narayanamurti & R. C. Dynes,** *Observation of Second Sound in Bismuth.* Phys. Rev. Lett. **28**, 1461 (1972).

[49] **B. D. Coleman & D. C. Newman,** *Implications of a nonlinearity in the theory of second sound in solids.* Phys. Rev. B **37**, 1492 (1988).

[50] **C. C. Ackerman & R. A. Guyer,** *Temperature Pulses in Dielectric Solids.* Ann. of Phys. **50**, 128 (1968).

Printed in the United States
By Bookmasters